CBI

The Confederation
of British Industry

THE

CBI ENVIRONMENTAL MANAGEMENT HANDBOOK

CHALLENGES FOR BUSINESS

Edited by Ruth Hillary

EARTHSCAN

Earthscan Publications Ltd
London • Sterling, VA

First published in the UK and USA in 2001
by Earthscan Publications Ltd

ISBN: 1 85383 637 0

Typesetting by JS Typesetting, Wellingborough, Northants
Printed and bound in the UK by Thanet Press, Margate, Kent
Cover design by Richard Reid

For a full list of publications please contact:

Earthscan Publications Ltd
120 Pentonville Road, London, N1 9JN, UK
Tel: +44 (0)20 7278 0433
Fax: +44 (0)20 7278 1142
Email: earthinfo@earthscan.co.uk
http://www.earthscan.co.uk

22883 Quicksilver Drive, Sterling, VA 20166-2012, USA

Earthscan is an editorially independent subsidiary of Kogan Page Ltd and publishes
in association with WWF-UK and the International Institute for Environment and
Development

A catalogue record for this book is available from the British Library

Library of Congress Cataloging-in-Publication Data

The CBI environmental management handbook : challenges for business / edited by
Ruth Hillary with Adam Jolly.
     p. cm.
  Includes index.
  ISBN 1-85383-637-0 (pbk.)
   1. Industrial management--Environmental aspects--Great Britain. 2. Social
responsibility of business--Great Britain. 3. Sustainable development--Great
Britain. 4. Environmental management--Great Britain. I. Hillary, Ruth. II. Jolly,
Adam.

HD30.255 .C36 2001
658.4'08—dc21

                          2001018786

This book is printed on elemental-chlorine-free paper

Contents

THE

CBI ENV
MANAG
HANDB

In August 2000 Wye College will merge with Imperial College of Science, Technology and Medicine which is also part of the University of London, and one of Europe's most prestigious education institutions.

WYE COLLEGE
UNIVERSITY
OF LONDON

For further details please contact:
External Programme
Wye College
University of London
Ashford Kent TN25 5AH UK
Tel: +44 (0)1233 813555 ext 280
Fax: +44 (0)1233 812138
Email: epadmin@wye.ac.uk
Web: www.wye.ac.uk/EP

UNIVERSITY OF LONDON
WYE COLLEGE

Postgraduate and Professional Development Programmes in:

- **Environmental Management**
- **Environmental Assessment**
- **Environment and Development**

Applicants have a unique opportunity to choose from a range of flexible study methods:

- **Distance Learning**
  - Study at home and in your own time, whilst remaining in your job
  - Use our interactive distance learning material and our newly developed on-line learning support system

- **Full-time study at Wye**
  - One-year MSc programmes

- **Mixed Mode Study**
  - Combine periods of distance learning at home with residential study at Wye

- **Short Courses**
  - For professional development at Wye College or a preferred location. Courses can be tailored to your individual or institutional needs

Enhancing access to continuing education worldwide

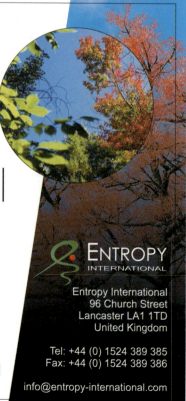

Foreword

Digby Jones,
Confederation of British Industry

There is an ever growing body of evidence to suggest that environmental issues are having a real impact on shareholder value. Management of these issues in addition to other, more traditional business risks is becoming an accepted aspect of effective business management. This new environmental business paradigm is being driven by a number of factors: government is encouraging business to be more environmentally responsible; consumers are demanding higher and higher environmental standards of the products and services they purchase and communities and non-governmental organisations (NGOs) are demanding a role in the business development process.

This is the first *CBI Environmental Management Handbook*. Expertly edited by Dr Ruth Hillary, it aims to provide today's busy business leaders with an introduction to the rapidly evolving environmental policy arena. It is a broad-ranging yet succinct guide to the vast number of environmental initiatives being undertaken by both UK and multinational companies. As well as outlining high profile initiatives such as emissions trading and adoption of accredited management systems the Handbook highlights some of the innovative approaches taken by business to address the environmental challenges that it faces.

The Handbook gives practical examples of the work UK companies are doing is areas such as producer responsibility, energy management, sustainable transport and corporate accountability. It is on issues such as these that the successful companies of the coming century will measure themselves, as well as by the more traditional indicators of success.

The contributors represent a broad spectrum of UK industry, each with new and innovative approaches to the evolving challenges and opportunities that the environment and sustainable development represent to both the UK and the world as a whole.

THE GREEN CONSULTANCY

More Profit and Less Tax
For Your Organisation Via Our
Award Winning
Energy Cost Reduction Services

The Green Consultancy offers a comprehensive range of Energy and Water Cost Reduction Services to enable your organisation to:

• make **substantial cost savings** – typically 15%-50%
• minimise your **Climate Change Levy** tax bill
• deal with the energy and water aspects of **Integrated Pollution Prevention and Control**
• reduce **environmental impact**.

Our aim is to exceed your expectations
All of our clients rate each aspect of our services as **equalled** or **exceeded expectations** and recommend us.

Expertise and experience that you rely on
Our energy management consultants each have decades of successful experience of energy management for all types of clients – including the following examples.

• educational establishments	• local authorities
• financial institutions	• manufacturers
• hospitals	• office blocks
• hotels	• plant nurseries
• leisure centres	• retailers

For more information on how to reduce your energy consumption and costs, please contact John Treble BA MSc at:

The Green Consultancy
Freepost RG219, Farnham, Surrey, GU10 5BR
Tel: (01420) 520900; Fax: (01420) 520901
Email: john@greenconsultancy.sagehost.co.uk

Why Energy Auditing?

Environment & Taxation

All organisations buy energy and water and all waste them to some degree – in fact, typically, **cost reductions of 15%-50%** can be easily achieved on a commercially attractive basis. Energy and water consumption have major environmental impacts that are well known. In particular, following the Kyoto Agreement, the UK Government has made international commitments to reduce energy consumption in order to reduce the greenhouse effect of gases such as carbon dioxide which is produced, together with other pollutants, when fossil fuels are burned on site or at power stations.

As part of its plan to fulfil these commitments, the Government is seeking the corresponding commitment of organisations to reduce their energy consumption. To encourage this an **Energy Tax**, The Climate Change Levy, is being introduced from April 2001. This tax is expected to increase the energy costs of most non-domestic consumers by approximately 11% on electricity and 19% on natural gas and solid fuel.

In addition, the EU directive on **Integrated Pollution Prevention and Control** requires organisations to minimise energy and water consumption as part of a more holistic approach to reducing the environmental impact of their total operation – rather than simply meeting standards for individual processes.

For many years the Government has been promoting Energy Efficiency Auditing as a means of quickly identifying cost-effective ways of reducing energy consumption.

A Green Consultancy Energy Audit For Your Organisation

Your audit will cover the entire energy and water usage of all buildings and processes and will consist of a comprehensive survey, followed by an investigation of the opportunities for reducing wastage, using existing facilities and new techniques or equipment, to achieve an overall cost reduction – whilst maintaining or improving comfort levels and/or production rates. Water is included because its use is partly interrelated with energy and because we can usually identify savings opportunities in that area too.

In addition, we will examine your existing tariff situation to ensure that you are paying as little as possible - but unlike tariff consultants we do not ask for any percentage of any savings achieved. For example, for a client spending £524,000pa on energy and water, we recently obtained an immediate electricity refund of more than double our consultancy fee. That audit also identified 30% annual cost savings with an overall simple payback period of only 5 months.

Yes, but do I really need one?

Before an audit, all clients are aware of some of the measures that they could take to save energy and water. What they lack is the total picture of all worthwhile opportunities with payback periods. Clients have often had bad experiences of implementing measures in an ad hoc way - having little or no idea of when they will get their money back - or having listened to the exaggerated claims of salesmen offering cost-saving equipment. Such salesmen are, in any case, extremely unlikely to know enough about your consumption patterns to make any reliable prediction of savings.

Your report

The recommendations in your report will be fully costed and graded to show:

• savings to be achieved without capital expenditure;

• further savings to be achieved with minimal expenditure and simple payback period within one year;

• further savings to be achieved with greater expenditure and with corresponding simple payback periods within a range specified by you.

Your report will be a powerful and easily understood management tool to enable you to start taking high quality decisions immediately. You will know that you are saving money as quickly as possible in the most cost effective way to suit the operation of your organisation.

Your report summary will be presented in a form ideal for the planning of investment and savings, taking into account any other relevant factors such as cash flow. Thus your report will not be a luxury but a valuable tool that will be of greater value to you the sooner you have it.

Exceeding Client Expectations

We aim to exceed clients' expectations and they all rate our services as **equalled** or **exceeded expectations** and recommend us. Typical comments include:

• *I was impressed by the depth and detail of the Audit, which was a thoroughly professional job.*
• *Amply demonstrated the benefit of using outside expertise that clearly exceeded our in-house ability.*
• *Of real benefit to the company in both short and long term.*
• *The detailed information generated, along with the expert advice, is not something that the company would have had the time to produce – or the necessary knowledge base across the range of activities assessed.*
• *The report findings more than pay back the investment.*
• *May I also comment on your consultant's excellent non-invasive approach which meant that as little of management time as possible was spent during the investigation.*
• *The savings indicated … will certainly fund the cost elements producing further savings not only in materials and service supply costs, but also in maintenance manpower.*
• *I am extremely grateful to … for their diligence and almost invisibility whilst carrying out the fact gathering in a most non-disruptive way.*
• *The appointment has been most worthwhile and was enjoyably productive.*
• *A greater awareness of the need for energy conservation is appearing in the organisation.*
• *I thought that energy consumption was managed frugally, but I was surprised by the level of potential savings identified over such a relatively short payback period.*
• *The recommendations enabled us to prioritise our plans.*
• *The consultant identified a serious omission on the part of our boiler maintenance contractor which would otherwise have gone undetected, with potentially worrying consequences.*
• *Clear guidance on replacement of energy relevant capital equipment.*
• *Other benefits include the identification of safety issues.*
• *Confidence of maintenance staff - which will greatly influence implementation.*
• *It has been a pleasure to report how useful your consultancy was (1 year afterwards).*
• *Cost reductions in the plant are ahead of expectations (2 years afterwards)*

Other Consultancy Services

The Green Consultancy's other energy management services include the following: Project Management, Plant Efficiency, Waste Recovery, Alternative Fuels, Training Seminars, Monitoring and Targeting (M&T), Tariff Analysis and Tendering.

 **University of Hertfordshire**

MSc/PgDip/PgCert
Environmental Management for Business
By Distance Learning in the Workplace

The Faculty of Natural Sciences and the Business School of the University of Hertfordshire offer a well established postgraduate programme, Environmental Management for Business, which can be studied by distance learning in the workplace over 24 months.

The aim of this postgraduate programme is to provide a basis for effective environmental management within a business setting.

Benefits to students

On completion of the course the student will be in a position to lead their organisation towards certification for environmental management standards such as ISO 14001 or EMAS. Distance learning saves time and money by allowing the student to remain in employment for the duration of the course. Regular short courses enable students to share experiences, meet their tutors and develop the knowledge and motivation to successfully meet the demands of the course. An 18 month course schedule, including all key target dates, is published in advance, allowing students to make more effective use of their time.

Who is the programme for?

This is a flexible modular programme of courses, which is designed for busy staff with responsibilities that include environmental management in business, industry, regulatory bodies and local government. Students have come from: Basildon District Council; Bass Breweries Limited; Eastern Generation; Engineering Employers Federation; EMI Music; Kier Construction; RAF Halton; One2One; Vauxhall Aftersales and many other organisations.

Course structure

Each course is studied over 10 to 14 weeks and comprises of 150 hours of study per module. The distance learning programme is supported by study packs, intensive two-day short courses (held on Fridays and Saturdays), case studies and workshops. Leading experts from various government departments, local authorities and environmental practitioners from business and industry contribute to the short courses.

The curriculum

* **Introduction** to Environmental Management Systems. * Environmental Review.
* Environmental Impacts Evaluation. * Implementing Environmental Management Systems (2 modules). * Integrating Environmental and Local Government Management.
* Environmental Auditing (2 modules). * Individual Project/Dissertation (4 modules).

For more information Contact: Dee Smith on 01707 286313 **OR** e-mail d.m.s.smith@herts.ac.uk **OR** * full details can be found on our web page: http//www.herts.ac.uk/natsci/Env/envman

The relationship between business and the environment is a dynamic one, and is one which business will have to grapple with more and more in the coming decades. This Handbook should act to facilitate the successful management of this relationship and provide sign-posting to businesses starting on the journey to sustainability.

Digby Jones is Director General of the Confederation of British Industry (CBI), the UK's leading employers' organisation.

Confederation of British Industry
Centre Point
103 New Oxford Street
London WC1A 1DU

tel: 020 7395 8247
fax: 020 7240 1578
website: www.cbi.org.uk

Introduction

Ruth Hillary,
Network for Environmental
Management and Auditing

The CBI Environmental Management Handbook explores the diverse and complex dimensions of environmental management as it applies to business. In 43 chapters, grouped in 11 themed sections, leading experts from the business community, government and its agencies and consultancies address the key macro and micro environmental issues facing companies today and in the future. This Handbook examines best practice and new approaches in timely and thought-provoking corporate case studies, critical policy analysis and commentary.

The first two sections consider macro issues with relevance to the company (micro) level. Section I considers sustainability and biodiversity; prominent strategic issues at the international and national levels (Walsh, DETR) that have great relevance to companies' strategic development as sustainable enterprises (Green, Wessex Water) and their management of biodiversity (Spray, Northumbrian Water) and access to biodiversity (O'Neill, GlaxoSmith Kline).

Section II addresses climate change with Chris Faye (ACBE) arguing for companies to address it as a strategic issue. John Browne (BP) illustrates how leadership has orientated BP to embrace the issue in a productive forward-thinking way. Further to this, the impact of the Kyoto targets on business (Murphy and Mundle, Shell International and PricewaterhouseCoopers) and permits for emission trading (Kyte, Powergen) are discussed.

Understanding the financial, legal and insurance ramifications of corporate environmental strategies are key to business success. Section III presents strong arguments for the often-disputed link between financial and corporate environmental performance (Edwards, Andersen Consulting) and the influence green funds have on corporate strategy (Millar, Jupiter Asset Management). Further to this, Paul Sheridan (McKenna and Co) shows that directors' personal liability under environmental law is real and a potential threat to managers. This, coupled with more stringent analysis by insurers, means that business needs to take account of its environmental risks and develop appropriate management strategies (Pritchard, Royal & SunAlliance).

Increasingly, products and their environmental impacts receive attention from all stakeholders including legislators. Section IV addresses this topic in four chapters. It considers integrated product policy (Atkin, Panasonic), producer responsibility (Jones, Biffa) and extended producer responsibility (Evans, Sony), with Rana Pant (Procter & Gamble) in the final chapter unpicking the environmental worth of the packaging waste regulations.

Mobility and communication are essential elements of both the old and new economies. Section V demonstrates the need for companies to effectively manage and improve their performance in transport as in the case of Pfizer (Elliot) and logistics management as shown by Tesco (Neville-Rolfe), while achieving bottom-line benefits. Telecommunications strategies are suggested as a way to introduce new working paradigms (Wood, BT) and Simon Berkeley and Tom Woollard (ERM) give a salutary warning to companies that neglect to fully understand the new communication model presented by the Internet.

Section VI considers the implementation of environmental management systems (EMSs) as techniques which both large and small companies can use to manage their environmental performance. Matthias Gleber (14000 & ONE Solutions) suggests formal EMSs such as ISO 14001 need to be broken into manageable stages for smaller firms because of their complexity and the resource limitations within these firms. The flexibility of Web-based EMSs may offer a viable solution to unwieldy systems (Gerstenfeld and Roberts, Entropy International). EMSs can be linked to existing sector standards such as Responsible Care (Richardson, Thomas Swan). In all cases, Chris Sheldon (BSI Quality) argues that the legitimacy of implemented EMSs is an issue for stakeholders.

The efficient management of energy is important to all companies and illustrated in case studies in Section VII. Increasingly, energy management is also driven by government incentives such as the climate change levy which is critically analysed by Judith Hackitt (CIA) and the search for economically viable forms of renewable energy generation, eg wind (Dinning, Scottish Power) and wave (Hill, BWEA) to reduce the dependency on fossil fuel generation. Uly Ma (EEBPP) takes a look at the important role of the workforce in energy-saving and waste-reduction measures.

The shift from UK to European Union (EU) legislation to reduce emission and prevent pollution is aptly revealed in Section VIII with the implications and analysis of the EU's air quality directives (Granville-West, CBI) and its Water Framework Directive (Taylor, AstraZeneca). The recently introduced integrated pollution and prevention control (IPPC) Directive is an example of how legislation now seeks to control pollution in an integrated way. Practical advice comes from Steve Bradley (British Sugar) on IPPC implementation. Stuart Stearn (Environment Agency) gives the regulator's perspective on IPPC.

Pollution abatement is often dependent on technological innovation. In Section IX, the trends and drives for clean technology are examined (Hilton, ECOTEC) and case studies illustrating cost savings (Gibson, ETBPP) and supply chain benefits (Crosbie, NIKE) are also presented.

Section X considers the pollution legacy of contaminated land as a real threat to companies if not managed proactively (Spray, Blue Circle). As highlighted by Tony Allen (Donne Mileham Planning), contaminated land and its liabilities are real and have been recently intensified with new regulations, as has the tightening role of planning restrictions on business activities (Thomas, S J Berwin).

This Handbook concludes with five chapters in Section XI on measuring and reporting environmental performance and stakeholder dialogue. John Elkington (SustainAbility) argues that these two difficult areas are often neglected by businesses but that they do this at their own peril. The importance of reporting has been recognised by some companies such as Anglian Water (Smith) and rewarded and encouraged by awards (Adams, ACCA). However, defining an effective measurement and benchmarking system is challenging (Urwin, Allied Domecq) but necessary suggests Sandra Palmer (United Utilities) to be able to facilitate stakeholder dialogue around real company environmental performance.

The CBI Environmental Management Handbook is essential reading for business charting the choppy waters of environmental management. It is not a definitive volume as it would be literally impossible to achieve this in such a vast field. The contributors, however, have provided examples of sound and thoughtful leadership and strategy on the environmental, economic and social issues that combine to make sustainable development. The Handbook benefits hugely from the lessons, pitfalls and achievements highlighted by real world practical experience of companies. It also presents some of the innovative management thinking needed to cut a path towards corporate sustainability.

Dr Ruth Hillary is founder of the Network for Environmental Management and Auditing (NEMA), UK Principal Expert to the International Organisation of Standards (ISO) TC 207 ES/1 ISO 14004 and Editor-in-Chief of *Corporate Environmental Strategy: International Journal of Corporate Sustainability*. Dr Hillary researched her PhD on 'The Eco-management and Audit Scheme (EMAS): Analysis of the Regulation, Implementation and Support' at Imperial College, University of London. She has been a consultant for the Department of Trade and Industry (DTI), the European Commission (EC) and many international companies. She has project managed many EU projects, including having been the UK National Co-ordinator for the EC's Euromanagement–

Environment pilot action on EMAS in smaller firms. She is widely published and is editor of *Small and Medium-sized Enterprises and the Environment* (Greenleaf) and *ISO 14001 Case Studies and Practical Experience* (Greenleaf). She is the series editor of the Business and the Environment Practitioner Series (Earthscan).

Network for Environmental Management and Auditing (NEMA)
174 Trellick Tower
Golborne Road
London W10 5UU

tel/fax: 020 8968 6950
mobile tel: 0771 471 8981
e-mail: rhillary@nema.demon.co.uk

Section I

Sustainability and Biodiversity

1
Government Sustainability Strategy and its Meaning for Business

Bernard Walsh,
Department of the Environment,
Transport and the Regions

While for some people the term sustainable development can be rather abstract, the basic concept is simple, namely integrating economic, social and environmental policies *'to ensure a better quality of life for everyone, now and for generations to come'*.

Sustainable business matters because the need for growth is as great as ever. But so too is the need to take account of the wider concerns at national and international level over the way in which business is conducted. By way of illustration, the UK currently uses approximately 10 tonnes of raw material to make 1 tonne of product and it is unsustainable to consume this at current rates. One of the key objectives is the prudent and effective use of resources. This is not simply a matter of resource depletion, as often alternatives can be found. It also concerns the cumulative impact of waste products on the environment, whether as polluting emissions or material going to landfill.

The backdrop to all this is an ever-growing world population – set to rise from 6 to 9 billion people by 2050; increasing pollution impacting on our lives – most notably through climate change; and concern for the social and ethical aspects of business practice. The challenge is to find new ways of meeting people's needs, expectations and aspirations. While all sectors of society have a role to play, for business it means producing goods and services in ways that are not only economic and commercially viable but that are also environmentally and socially acceptable.

A key challenge for government is to find the right mix of regulatory, economic and other instruments, including voluntary initiatives by business, and the provision of information, to encourage and facilitate progress and ensure that everyone – individuals, local communities, voluntary bodies, international organisations and – critically – businesses play their part. That is why in May 1999 the government published its sustainable development strategy 'A Better Quality of Life', which emphasised four themes:

- social progress that recognises the needs of everyone;
- effective protection of the environment;
- prudent use of natural resources;
- maintenance of high and stable levels of economic growth and employment.

Government will use the strategy, together with the 'Climate Change Programme' and the associated indicators in 'Quality of Life Counts' as the framework to guide its policies and judge future progress.

Regulation will continue to underpin basic minimum requirements and where performance is not acceptable. While the aim is for better, not necessarily more, regulation, the standards set by regulation are likely to rise. Business needs to be aware of this and to plan for what is on the horizon.

Over time government intends to reform the tax system: to shift the burden of tax from 'goods' to 'bads' to encourage innovation in meeting higher standards and a more dynamic economy and cleaner environment.

For business the challenge is wider corporate social responsibility – about a continuing commitment to behave ethically and contribute to economic development while improving the quality of life for its employees, the community and society. These are sentiments being taken up around the world and echoed for example by the UN's 'Global Compact', aspects of which are being reflected closer to home in specific ways such as:

- the need for sound internal risk management as advocated in the 'Turnbull Report' published in September 1999 by the Institute of Chartered Accountants;
- new regulations made under the Pensions Act now require occupational funds to report the extent to which, among other things, environmental considerations are taken into account in their investments;
- the Company Law Review is consulting on proposals to provide for reporting performance on environmental, social, ethical and reputational issues in company reports;
- interest in aspects other than financial performance is also being stimulated by the emergence of various business indices and surveys – BiE's Index of Corporate Engagement, the PIRC Survey, Dow Jones Sustainability Group Index, etc.

To respond to all these challenges and pressures business now has access to a growing variety of tools, including:

- **management systems** – ones covering quality, environment, health and safety, investing in people, as well as social and ethical issues for which there are national and international standards (eg ISO 14001, EMAS, AA 1000, SA 8000, BS 8800, OHSAS 18001);
- **key performance indicators** – such as the government's sustainable development indicators, those in ISO 14031 and the eco-efficiency indicators developed by the World Business Council for Sustainable Development;
- **product management** – standards for product stewardship and consumer information (eg ISO 14021, the government's Green Claims Code);
- **environmental reporting** – making full use of the various guides currently available such as the 'Global Reporting Initiatives' and the government's reporting guides for greenhouse gas emissions, waste, water.

Government is also supporting the development of the next generation of sustainable management systems – Sigma – as well as ways of encouraging smaller businesses to embrace environmental management – Acorn.

To encourage organisations to set targets and in particular to improve performance in greenhouse gas emissions, waste and water – all matters of national and international concern – the government has relaunched 'Make a Corporate Commitment' ('MACC2').

Sustainable business requires innovation and creativity on an unprecedented scale involving the adoption of new processes, new products and new ways of doing business. To that end business can access the government's 'Best Practice Programmes' and the 'Sustainable Technologies Initiative' being introduced to help businesses develop technologies to incorporate sustainability into their products and processes from the design stage.

The government's strategy also identifies a role for trade associations by developing sectoral sustainability strategies to provide a framework for concerted, co-ordinated action in the light of economic, environmental and social pressures over the next 10 years or so, including targets for improving performance and reporting on progress. There is also scope for working with the market, for linking voluntary effort with the regulatory and economic instruments, often called 'market transformation', to provide additional incentives to encourage and underpin improved performance.

The aim is to encourage more sustainable business. Many companies already know what is required and some are ahead of the game. For those still getting to grips and catching up there are tools available to help and additional ones under development. The government is determined to enhance the support and incentives it can provide and will monitor progress, adjusting the policy mix accordingly.

Some practical steps business can take

- **Make a commitment** to manage and improve your impacts – resources, energy and water, waste, transport, emissions, etc – using appropriate management systems ISO 14001 (*www.iso.ch*) and EMAS (*www.emas.org.uk*), and by signing up to the government's relaunched Make a Corporate Commitment campaign (*www.macc2.org.uk*).
- **Be 'best in class'** – explore the scope for greater resource efficiency, benchmarking performance (eg *www.dti.gov.uk/ukeei*), making fullest use of the government's Best Practice Programmes (*helpline* 0800 585794) and Sustainable Technologies Initiative (01296 337 165).
- **Embrace the principles of producer responsibility** by taking account of the different aspects from 'cradle to grave' in supplying products and services, working towards greater recycling and recyclability, and considering all the implications and opportunities at the design stage.
- **Attend to social responsibilities** though good employer practices by, for example, encouraging fairness at work; helping staff to develop their skills; introducing green transport plans; being a good neighbour, responsive to the local community; and being an ethical trader.
- **Communicate with stakeholders** – reporting on environmental performance against meaningful targets, using relevant certification schemes and making product declarations that are legal, decent, honest and truthful.
- **Work with others** either through the supply chain, specifying what you want and helping others to comply, or as part of concerted sectoral action to help improve overall performance, safeguarding yours as well as theirs.

Bernard Walsh heads a team in the Environment, Business and Consumers division within the DETR covering different voluntary approaches to sustainable business.

Department of the Environment, Transport and the Regions (DETR)
Zone 6/D9
123 Victoria Street
London SW1E 6DE

tel: 020 7944 6572
fax: 020 7944 6559
e-mail: bernie_walsh@detr.gsi.gov.uk

2

What is the Sustainable Enterprise?

Dan Green,
Wessex Water

Wessex Water's activities are in constant interaction with the water cycle, and the environment is indivisible from everything we do. We have always appreciated our close interactions with the environment; nowadays we take a broader view, reflected in our 1997 commitment to becoming a *sustainable* operation. For Wessex Water, sustainability means providing our customers with high quality products and services in a way that does not damage – and may even enhance – the environment we use every day of our working lives. This involves thinking and acting beyond the traditional business measures of profit, and using a holistic approach to wealth creation, giving as much importance to human and social issues as to product delivery and the natural environment. We now consider our resources using a framework developed by the sustainability charity Forum for the Future, which focuses on five types of 'capital' – ecological, human (employees), social/organisational, manufactured and financial.

Our goal of a sustainability approach did not emerge from a crisis, but through consideration of the role of four equally weighted 'stakeholders' that underpin our approach to business:

- **Shareholders** – demonstrating responsibility in actions is attractive to investors. Our sole shareholder is our parent company Azurix. We believe that demonstrating diligence in economic, environmental and social senses can help expand Azurix's business worldwide.
- **Customers** – we have a duty to serve our customers. Our reputation is also enhanced if customers realise that we are interested in more than the economic bottom line.

- **Employees** – a healthy, motivated workforce that shares the company's ideals is self-evident as a benefit to business.
- **Environment** – our environmental dimension is closely regulated, but we also recognise the good links forged with the public and interest groups around environmental issues and the potential for greater efficiency that those links can provide.

In practical terms there are several facets to the sustainability of a water company. The following are a cross section of the sustainability issues with which we are engaged on a day-to-day basis:

- We cannot compromise the safety of drinking water.
  Action – as well as testing around 60,000 samples of drinking water each year we are encouraging farmers in targeted areas near vulnerable water sources to convert to organic farming or reduce nitrate inputs.
- We can work hard in our operations and with our customers to improve efficient water use.
  Action – a leakage reduction programme; voluntary, free metering; domestic water audits and cistern devices contribute to water efficiency.
- We aim to provide full treatment and appropriate disinfection for all coastal discharges that affect recreational or shellfishery waters.
  Action – we are installing membrane and ultraviolet treatment at various locations around the coast.
- We view biogas and biosolids (sewage sludge) as resources, rather than waste, particularly for their calorific and nutritional value.
  Action – we use biogas to generate renewable electricity and dry sludge to produce Biogran, a soil conditioner used in land reclamation and agriculture.
- Extending treatment to meet the higher standards demanded of us means that more and more energy is required.
  Action – we are looking for sources of sustainable energy and have a long-term aim of deriving 100 per cent of our energy from renewable sources. We have pushed forward the development of a global warming indicator as a tool for reporting on environmental performance in the corporate sector, and have developed a 'slimmed-down' tool for life cycle assessment, used by engineers while considering new options for infrastructure.
- Our region is of great importance for its biodiversity.
  Action – our biodiversity action plan promotes threatened species and habitats in our region, both through our own work and through partnership with local expertise.
- The well-being of our staff is of immense importance – training, personal development and health and safety are therefore central to sustainability.
  Action – in-house health and safety professionals monitor the varied workplaces in the company. Training covers this issue and also helps to extend our employees' development across a range of skills.

- Water has a social agenda that has been neglected in company reporting, particularly reports that focus on the short-term financial returns of investment. Successful organisations benefit from healthy communities and we constantly aim to ensure that charging systems for water reflect fairness. *Action* – low user tariffs, monitoring of service standards and customer perception, proactive community links and information provision are all extending the social dimension of the sustainability of our company.

The water industry is already relatively mature in addressing sustainability. Biosolids are no longer viewed as a 'waste' but as a resource. Many players within the sector have recognised the biodiversity importance of their land and are tackling this issue in a strategic way. The social dimension of water services is recognised and there is lively debate on where society's aspirations and the needs of the environment match and where they diverge. There is recognition that there is a major international dimension to meeting the basic social necessity of water. However, there are 'crunchy' challenges at the heart of our core business that we will face for many years. How do we reconcile increasing energy consumption that is linked to higher treatment standards? What other forms of waste – in physical and social terms – are we guilty of generating?

We have not found immediate, easy answers. Instead we see that working towards sustainability is a dynamic process that must be viewed in the context of long-term issues. We have taken the first steps, and look forward to the story unrolling over the next few decades.

Dan Green is Wessex Water's sustainability co-ordinator, developing strategy and policy on a range of issues, including carbon management, sustainable agriculture, biodiversity, sustainable management systems and reporting. Trained as a geographer (BA at Durham and Aix-Marseille and PhD at the Countryside and Community Research Unit in Cheltenham), Dan entered the world of sustainability and business via Forum for the Future's scholarship programme.

e-mail: dan.green@wessexwater.co.uk

Wessex Water is the regional water and sewage business which serves an area of the south west of England covering 10,000 square km, including Dorset, Somerset, Bristol, Bath, most of Wiltshire and parts of Gloucestershire, Hampshire and Devon. Wessex Water supplies around 380 million litres of water per day to homes and industry and serves 1.2 million people in 500,000 properties. Wessex Water treats sewage from around 2.5 million people.

In the early 1990s the company declared that the environment, customers, employees and shareholders have an equal footing as 'stakeholders' in its business. Shortly after, following its first work with Forum for the Future, Wessex Water committed itself to becoming a sustainable operation.

10   Sustainability and Biodiversity

Wessex Water
Claverton Down Road
Bath BA2 7WW

tel: 01225 526 000
website: www.wessexwater.co.uk

3

The Business of Biodiversity

Chris Spray,
Northumbrian Water

Introduction

Industry as a whole has been slow to respond to the challenges of biodiversity, and the conservation of biological diversity (or variety of life) has rarely been a serious agenda item for the board room. However, conserving the UK's biodiversity is an essential requirement for sustainable development and an increasing priority of both government and industry. Environmental liability, climate change, corporate governance, stakeholder pressures, environmental reporting and competitive positioning are all focusing attention on the natural environment and the way we interact with it.

The government published the 'UK Biodiversity Action Plan' (BAP) in 1994, since when the UK Round Table on Sustainable Development and the Department of the Environment (now the Department of the Environment, Transport and the Regions (DETR)) have published a guide to Business and Biodiversity, and Case Studies in Business and Biodiversity. Together with published industry examples of best practice – from the water, minerals and petroleum sectors – these provide a good guide to effective biodiversity management for business.

Why business should be interested in biodiversity

At one time concern for biodiversity would have been seen as little more than altruism, of no more particular relevance to business than any other type of charitable activity. However the real value of involvement in biodiversity is much more fundamental:

- **Biodiversity is a key element of environmental reporting**. As more and more companies produce environmental performance reports so the need to address biodiversity in a structured manner increases. Furthermore the government threatens to make environmental reporting mandatory.
- **Customers place environmental issues high on their agenda.** Customer surveys, across the water industry for example, have shown that customers have a strong interest in environmental matters. Membership of conservation organisations directly involved with biodiversity, such as the Royal Society for the Protection of Birds (RSPB) (over 1 million members), is vast and continues to grow.
- **Environmental issues are an increasingly important part of the contract tender process when bidding for new business.** While not yet true for smaller companies, this is becoming more prevalent among larger companies, and can be expected to transfer down the supply chain in time. In many instances biodiversity is one of the areas about which information is required.
- **Corporate governance, employee involvement and community responsibility are increasingly important.** In this respect biodiversity can be seen as one of the areas to consider, and can be an important part of a company's local community relationships, education programmes and employee engagement.
- **Environmental liability and risk are driving board room agendas with increasing urgency.** The potential impact, for example of accidental pollution on the biodiversity of a European conservation site, may carry huge penalties in the future.
- **Biodiversity represents the wealth of raw materials for some business sectors.** Many plants have direct commercial value, as well as ecological functions in their own right.
- **Biodiversity can be an important element of company image and brand.** The list of companies with an animal or plant as a brand identifier is long, as is the number and extent of corporate sponsor of wildlife.

Identifying the main challenges for your company

For some businesses, such as landowning or extractive companies, the challenges will be obvious, and can largely be met by the manner in which their land is managed. For others the connection is much less obvious and may require more creative thinking – for example where and how their main raw materials are resourced and what impact this has on the local biodiversity. Four key areas can be identified:

- land management;
- operations and investment practices;

- funding and sponsorship;
- community and workforce involvement.

The national and regional context

Companies need to place their biodiversity programme in the context of the government's BAP and, more importantly, their regional and local BAPs. For smaller companies particularly reference to their local BAP will enable them to see where best they can target their effort and to whom they should turn for advice and support. Larger businesses, however, should not work in isolation – they have much to gain from partnership building, promotions and reporting.

Biodiversity in land management

Managing land for biodiversity enhancement involves five stages:

- **Habitat and species surveys** – it is often best to use the local Wildlife Trust ecological consultancy to carry this out, thus building relationships and tapping into local expertise.
- **Audit** – assessing the distribution and abundance of key species and habitats of conservation importance on your lands.
- **Identifying BAP priorities** – in conjunction with external partners, to identify which of the UK BAP's key habitats and species are represented on your land, and how important they are locally.
- **Producing biodiversity action plans** – for each key habitat and species; to include information on current status locally and on your lands, management practices and threats, and targets.
- **Implementation and reporting** – action, monitoring and reporting; this latter to include reporting internally, to the public and through your local BAP into the main UK process.

Biodiversity in operations and capital investment projects

Impacts can be viewed as either direct or indirect, and are best approached through an Environmental Management System (EMS) such as ISO14001.

Direct impacts

Northumbrian Water uses an environmental screening process for each new capital investment project. The system involves reference to a GIS on which all local conservation sites are recorded, and use of a 'biodiversity matrix' that identifies potential impacts of key operations and activities on BAP habitats and species. By subjecting operations and new investment to this process damage to biodiversity is avoided and opportunities for enhancement can be identified.

Indirect impacts

Purchasing policies, waste minimisation, energy management and other activities can have an indirect impact on biodiversity, though the link may not be readily apparent and the impact not local. Influencing the impact is best done via targeted projects as part of an environmental improvement programme, such as a water leak reduction programme.

Both indirect and direct impacts can also be approached via risk analysis – Northumbrian Water has developed an Environmental and Biodiversity Risk Assessment Register (EBRAR) to cover this aspect at over 250 operational sites.

Biodiversity and sponsorship

The biodiversity process provides some great opportunities for sponsorship through the championing of species BAPs either nationally or locally. Nationally companies such as ICI, the Co-op Bank, Northumbrian Water and Tesco have signed up as BAP champions for species as diverse as the large blue butterfly, bittern, round-mouthed whorl snail and skylark respectively. Along with this have come partnership working, publicity and effective targeting of charitable giving where it achieves best value. Locally businesses can get involved with similar schemes with their respective local BAPs.

Championing biodiversity with the workforce and in the wider community

Biodiversity action provides plenty of opportunity for workforce, customer, supply chain and community involvement. Sainsbury's, for example, has sponsored farm BAPs to influence its suppliers' activities, while Mileta Tog 24, national champion for the stag beetle, has produced information for customers at its outdoor clothes shops. To raise awareness of biodiversity issues, Essex & Suffolk Water arranged for its staff to work on a local RSPB reserve as part of their internal training.

Biodiversity indicators, monitoring and reporting

Monitoring BAPs and reporting on progress is important. While this can be achieved through an EMS, the results need to be communicated internally, to the local BAP organisers and to the general public. Local BAP organisers have developed a number of reporting methodologies, and it is vital they receive the information.

While the government has produced headline indicators of sustainable development, one of which covers wild bird populations, there are as yet no accepted indicators for biodiversity itself. The water industry produced the first sector-wide report on environmental indicators, and Northumbrian Water has taken this further with the RSPB, British Trust for Ornithology and the DETR, to produce regional biodiversity indicators covering specific wetland habitats.

The use of websites to report on biodiversity is another increasingly attractive method, either on their own or as an element of environmental reports.

Where to go for more information

- *Biodiversity: The UK Action Plan* (1994) – published by HMSO, PO Box 276, London SW8 5DT.
- *Business and Biodiversity: A Guide to Integrating Biodiversity into Environmental Management Systems* (1998) – published by Earthwatch, 57 Woodstock Road, Oxford OX2 6HJ, on behalf of the UK Round Table on Sustainable Development.
- *Case Studies in Business and Biodiversity* (2000) – published by Earthwatch, 57 Woodstock Road, Oxford OX2 6HJ, on behalf of the DETR.
- *RSPB Good Practice Guide for the UK Water Industry: Meeting the Biodiversity Challenge* (1998) – published by the RSPB, The Lodge, Sandy, Bedfordshire SG19 2DL.
- *Biodiversity: The UK Petroleum Industry Association* (1999) – published by the UK Petroleum Industry Association, 9 Kingsway, London WC2B 6XF.
- *Biodiversity and Minerals: Extracting the Benefits for Wildlife* (1999) – available from English Nature Enquiry Service, Northminster House, Peterborough PE1 1UA.
- *Northumbrian Water Biodiversity Strategy* (1998) – available from Northumbrian Water, Environment Department, Abbey Road, Durham DH1 5FJ or www.nwl.co.uk/biodiversity.

Useful websites

- **UK Biodiversity Group** – www.jncc.gov.uk/ukbg
- **Department of the Environment Biodiversity Secretariat** – www.jncc.gov.uk/ukbg/
- **National Biodiversity Network** – www.nbn.org.uk
- **Northumbrian Water Biodiversity Strategy** – www.nwl.co.uk/biodiversity

Dr Chris Spray, MBE, is the Environment Director with Northumbrian Water Group, based in Durham. He joined the water industry in 1984, after 10 years in academic research, latterly as a research fellow in zoology at Aberdeen University. A wetland ecologist and waterfowl biologist by training, he worked for Anglian Water Authority, then the National Rivers Authority, before joining Northumbrian Water in 1991. He is a member of council of the RSPB, a director of the British Trust for Ornithology, Chairman of Tweed Forum and a past director of the River Restoration Centre. He was appointed to the Government's Advisory Committee on Releases to the Environment for his expertise on biodiversity, and is on the Biodiversity Secretariat's England Local Group. He was made an MBE in the Queen's birthday honours in June 2000, for services to environmental improvement and conservation in the water industry. He represents the CBI on the North East Round Table on Sustainable Development.

Northumbrian Water
Abbey Road
Pity Me
Durham DH1 5FJ

tel: 0191 301 6758
fax: 0191 3016578
e-mail: chris.spray@nwl.co.uk

Northumbrian Water Ltd provides water and sewerage services to 2.6 million people and over 69,000 businesses in the north east of England, and water services only to a further 1.7 million customers in the south east. Formed by the merger in 2000 of Northumbian Water and Essex & Suffolk Water, it has a turnover of around £480 million and employs nearly 2,200 people. Among its key assets are 37 impounding reservoirs and 71 water treatment works, as well as over 400 sewage treatment works. It is part of Suez Lyonnaise des Eaux, which is one of the largest water utility companies in the world. One of the first companies in the country to achieve registration to the then BS7750 Environmental Management System, Northumbrian Water now has registration to ISO 14000 or ISO 9000 throughout. It launched its Biodiversity Strategy in 1998 and is working with English Nature and other partners on a range of Biodiversity initiatives, notably as national champion for the otter, black grouse, roseate tern and round-mouthed whorl snail.

Northumbrian Water
Environment Department
Abbey Road
Pity Me
Durham DH1 5FJ

tel: 0191 383 2222
fax: 0191 384 1920
e-mail: communications@nwl.co.uk
website: www.nwl.co.uk

4

Accessing Biodiversity for Industry

Melanie O'Neill,
GlaxoSmithKline*

Biologically diverse organisms ('biodiversity') have provided us with a vast array of raw materials that finds commercial exploitation especially in the pharmaceutical, agrochemical, health care, horticultural and cosmetic sectors of industry. The potential to yield valuable new products is the basis of an intense interest in accessing biodiversity for research and development.

Through previously bitter experiences which have left them feeling that their natural resources have been plundered for little or no economic gain locally, some countries, especially those of the developing world that hold the richest sources of biodiversity, have become reluctant or even refuse to allow species to be evaluated by foreign groups, particularly those which are commercially driven. Industry needs to understand the impact that unauthorised and unrestrained removal of natural materials from their indigenous habitats can have on the ecology and economy of a country. In seeking access to biodiversity industry must collaborate with organisations that possess the expertise and authority to obtain such materials.

The conservation of biodiversity, the sustainable use of its components and the fair and equitable sharing of the benefits arising from the use of genetic resources are the goals of the Convention on Biodiversity (CBD) that was initiated at the United Nations Conference on Environment and Development (known as the Earth Summit) held in Rio de Janiero in 1992. The concepts behind the CBD are the sovereignty of states over their genetic resources, obligations on nations to facilitate access and that contracting parties will

---

* This chapter was written while the author was Head of the Department of Compound Diversity and Biodiversity Officer at Glaxo Wellcome.

establish with the source nation measures for benefit sharing in the event of commercial utilisation.

The CBD has been ratified by 174 countries to date and awareness in industry of the concepts behind it continues to grow. Source countries are obliged to develop legislation to cover procedures by which access is granted and to develop mechanisms by which benefits can be used to support conservation and legislation. Industry seeking to participate in bioprospecting needs to define which samples are included or excluded from collection, to monitor compliance with this and to define benefit sharing arrangements covering sample collection costs, milestone payments, royalties, technology transfer and information exchange. These arrangements should set out benefit ceilings and indicate the expected level of contribution towards the developed compound or to the marketed product for the ceilings to be paid.

Some key issues have to be faced by industry wishing to access biodiversity and seeking to work within the framework of the CBD. Legislation to define access is not yet in place in most countries; in some that have defined legal procedures a high level of bureaucracy has to be grappled with before access will be granted. Industry needs to ensure that the organisations which will undertake collection are sufficiently skilled in taxonomy to ensure that collection of endangered species is avoided and there is an expectation that industry will introduce measures to monitor compliance among its suppliers.

The obligations on and to *ex situ* organism collections, such as botanic gardens and microbial culture collections, are unclear and the possibility exists that these organisations will need to deal with ownership claims from source countries in the event that a product of biodiversity in an *ex situ* collection turns out to have commercial value. In order to avoid a situation which might hinder or close an R&D programme midway, it is crucial for industry to clarify at the outset that the curators of such collections have the authority to supply samples for evaluation.

The benefit sharing obligations agreed by industry under the terms of contract with the providers of samples are likely to relate to any derivatives produced directly or indirectly from compounds isolated from biodiversity samples. It is therefore a clear requirement of recipients of samples that they have effective tracking systems which can log the progress of such compounds and their derivatives. Derivatives of biodiversity would also include genetic-ally modified organisms generated using DNA derived from wild strains.

In some cases biodiversity samples are information rich. For example, named plant species may be used by local people in traditional medicine to treat certain diseases. This information may be passed to industry to be used in the discovery of new and valuable medicines. Benefit sharing arrangements must take into account the potential added value provided by the information supplied with a sample.

The procedures required of industry before access to biodiversity is granted can seem tortuously bureaucratic at best and unacceptably labour intensive at worst. A number of organisations, mainly non-governmental organisations

(NGOs), have sprung up to help facilitate the process in various countries. Working with these organisations can considerably reduce the workload on industrial groups seeking to establish partnerships to unlock the remarkable potential of nature in giving clues to the invention of new commercial products.

For an in-depth review of industry and the Convention on Biodiversity, the reader is recommended to The Commercial Use of Biodiversity: Access to Genetic Resources and Benefit-sharing *by Kerry ten Kate and Sarah Laird (1999), Earthscan. ISBN 1 85383 334 7.*

Melanie O'Neill is Vice President of Information Management at GlaxoSmithKline (GSK). At the time of writing, she was Head of the Department of Compound Diversity and Biodiversity Officer at Glaxo Wellcome, responsible for establishing collaborations with organisations from various countries to enable Glaxo Wellcome to access natural source materials for research. She was instrumental in defining Glaxo Wellcome's 'Policy for the Acquisition of Natural Source Materials' that was introduced in 1992.

Melanie has a PhD in Pharmacognosy from the University of London. She is author of over 50 research papers in the field of natural products science and editor of the journal *Phytotherapy Research*.

GlaxoSmithKline plc
Research and Development
Gunnels Wood Road
Stevenage SG1 2NY

tel: 01438 745 745
fax: 01438 764 502

GlaxoSmithKline (GSK) was formed in December 2000, following the merger between Glaxo Wellcome, and SmithKline Beecham. One of the world's premier health care companies, GSK is a research-based pharmaceutical organisation whose people are committed to fighting disease by bringing innovative medicines and services to patients and the health care providers who serve them throughout the world.

GSK is headquartered in the UK and employs approximately 100,000 people in 76 operating companies worldwide, with manufacturing sites in 41 countries. It supplies product to over 150 markets. Prior to the merger, Glaxo Wellcome Research and Development worked with small quantities of natural source materials to discover bio-active principles. The focus of natural product lead discovery at GSK will differ slightly.

GlaxoSmithKline plc
Stockley Park West
Uxbridge UB11 1BT

tel: 020 8990 9000
fax: 020 8990 4321
website: http://corp.gsk.com/

Section II
Climate Change

5

Climate Change: A Strategic Issue for Business

Chris Faye,
Advisory Committee on
Business and the Environment

Climate change has the potential to impose enormous costs on society and the economy, even though uncertainties remain on timing and the scale of the impact. Business leaders increasingly believe that this challenge can be met, providing there is a firm baseline of international agreement that builds on the achievements of Kyoto.

By responding now business can manage and control the process in a cost-effective way, but the approach must be long term, and flexible enough to be adjusted in the light of practical experience. A policy framework is needed that will stimulate innovation, accelerate the take-up of new technology, harness commercial opportunity and provide the necessary incentives to achieve behavioural change across all sectors. ACBE has stressed also that such a framework should not damage the competitiveness of UK business and that the burden of change must not fall on industry alone – every consumer, homeowner and transport user can also play their part.

Business and energy use

The essence of the climate change challenge is to develop products and services with less reliance on carbon-based energy. It is estimated that there is substantial scope for further CO_2 savings in all business sectors of around 15–20 per cent – or around 10 million tonnes of carbon. Most of this can be saved cost-effectively: there are bottom line business benefits.

Many of the largest energy users have already implemented fuel savings, but a lot of businesses do not see energy use and fossil fuel consumption as core issues. There has been little financial incentive for change; the fall of energy prices since deregulation has enabled companies to reduce energy costs without focusing on energy use. Yet it is absolutely vital that all businesses are aware of what their energy usage is and how it can be reduced. This means generating awareness and responsibility in every member of the team.

ACBE has recommended that every business should measure and report on its CO_2 consumption using simple measures such as fuel bill information and following the Greenhouse Gas Reporting Guidelines that have been produced by business and government working together. Businesses should also be seeking information on fuel breakdown from energy suppliers and reporting on energy derived from combined heat and power and renewables.

Economic instruments

ACBE has welcomed the government's proposed climate change levy but believes that it needs significant modification. In particular, it should focus on carbon as far as practical, must not damage energy intensive sectors and should create substantially increased incentives for low carbon technology take-up and innovation. ACBE recommends that the levy should be neutral on its impact on business and should be recycled with the priority objective of reducing carbon emissions. A proportion (at least 20 per cent initially rising to 50 per cent) should be available for recycling and investment in independent, business-administered carbon trusts to support specific carbon reducing measures. The levy should also support a further allocation for investment tax credits or grants for leading edge low carbon technology.

ACBE also recommends that the UK government should fully support the rapid development of the flexible mechanisms of the Kyoto Protocol: joint implementation, the clean development mechanism and the provision for international emissions trading. ACBE believes that these mechanisms will allow the market to seek out the lowest cost options for carbon abatement and offer foreign market opportunities for British business. ACBE stresses, however, that transaction costs should be kept low and that emission reductions achieved should be fully tradable.

Carbon emissions trading will have a vital role to play in tackling climate change. The concept is a simple one: businesses can buy and sell carbon emissions to meet an emissions target. If a company produces below its emissions quota in the relevant timescale it will be able to sell the surplus quota to other firms that have proved unable to meet their targets. Carbon emissions trading should prove both cost-effective and a suitably flexible avenue for business to meet environmental targets. It will also offer the potential to encourage a wider range of businesses to commit to binding targets than those currently involved in the climate change levy negotiated agreements. Reduced levy rates should be available for those companies that take action through the approved schemes. The scheme will also keep the UK in the forefront of international emissions trading.

A CBI/ACBE Emissions Trading Group has been working with the government to develop outline proposals for a UK greenhouse gas emissions trading scheme. This will be open to all companies operating in the UK and will run alongside the climate change levy. Twenty-five leading UK companies are willing to support further development of the scheme in order that it can run from April 2001.

The sectoral approach

The climate change challenge can appear daunting for the individual company. The most practical way to engage many businesses is therefore likely to be on a sectoral basis. Sectoral organisations can help to identify options and ensure that business is fully involved in the development of policy. Heavy industry alone cannot solve all the problems for business. The food and drink and tobacco industries, for example, are also significant producers of CO_2, as are retail and commercial office sectors. Indeed it is estimated that improving the office use of energy could bring savings to business of around £700 million per annum.

Each business sector will be affected by climate change in a different way. Each sector needs to identify the likely impacts of changing weather patterns (eg water supplies) so that issues of particular concern can be addressed strategically.

Climate change levy negotiations between the DETR and the 10 principal energy intensive sector associations are currently taking place. The government recognises the need for special consideration to be paid to these sectors given their high energy usage, and it is intended to set significantly lower rates for sectors that can agree targets for improving their energy efficiency which meet the government's criteria. Setting targets at the sector level allows for a more flexible approach to be taken to meeting these targets.

Technology and research

Business has an important contribution to make in developing technology that will help to meet the UK's targets, and the government needs to ensure that funding is available for well-targeted climate change-related product development and research. ACBE has pointed to the US Climate Change Initiative as a good example of such support. In addition, the Committee recommends that a business-led Climate Technology Centre be set up. This will co-ordinate research and assist in the commercial exploitation of new technology, such as small-scale power generation, sensors, lighting, motor vehicles, renewables and the application of IT.

Lower carbon energy sources and renewables

In the shorter term the UK's initial climate change programme should encourage the greater take-up of combined heat and power (CHP) and the expansion of renewable energy. Renewables and 'green energy' generation, including CHP, are at a critical stage of development in the UK. There is substantial scope to boost energy generated from renewables to reach a 10 per cent target by 2010, provided both government and business increase their support for this sector now.

Conclusion

Each of these measures will contribute to the UK's target for reducing CO_2 emissions but the burden cannot be borne only by business. We need to see programmes and measures that will address the substantial opportunities available in transport and in the public and domestic sectors. It will require strategic direction, thorough consultation, business creativity, and innovative market solutions. It will also require a long-term partnership between business and government but there are substantial opportunities for business if we can generate the necessary partnership on programmes and measures. Business has a key role to play in building a more sustainable future and it is very much up to each of us how effectively and successfully we choose to respond.

Chris Fay, formerly chairman and chief executive of Shell UK Ltd, became Chairman of ACBE in June 1999. He has worked at the leading edge of business environmental and societal responsibility, initiating and producing the first externally verified Environmental Report in the Shell Group in 1997 and in 1998 furthering this with the first all-encompassing Report to Society. He is currently non-executive director of BAA plc, Anglo-American plc, Stena Drilling Ltd, deputy chairman of Stena International bv and chairman of the British Committee DnV.

ACBE Secretariat
Environment, Business and Consumers Division
Department of the Environment, Transport and the Regions
Zone 6/E12
Ashdown House (V)
123 Victoria Street
London SW1E 6DE

tel: 020 7890 6278
fax: 020 7890 6559
e-mail: neil_riddell@detr.gsi.gov.uk
DETR website: www.detr.gov.uk
DTI website: www.dti.gov.uk

The **Advisory Committee on Business and the Environment** (ACBE) was established in May 1991 in response to a commitment in the 1990 Environment White Paper, *This Common Inheritance*. The Committee provides for dialogue between government and business on environmental issues and aims to help mobilise the business community in demonstrating good environmental practice and management. The members are jointly appointed by the Deputy Prime Minister and the Secretary of State for Trade and Industry and serve in a personal capacity. ACBE has produced eight progress reports on its work to date. A special report, *Climate Change: A Strategic Issue for Business*, was presented to the Prime Minister in March 1998. This report's findings on economic instruments provided a basis for the review of this topic led by Lord Marshall. This has been followed by the government's proposals for a climate change levy and, more recently, for the development of a business initiative on carbon emissions trading (led by a ACBE/CBI Emissions Trading Group). ACBE has recently issued a report on carbon trusts and low carbon technologies. ACBE has also considered how business might contribute to the development of the flexible mechanisms outlined in the Kyoto Protocol to aid the reduction of greenhouse gas emissions and the achievement of sustainable development.

6

Climate Change:
A Corporate View

John Browne,

BP

The environment has been an issue of international public concern for more than three decades. During this time people have increasingly looked to business to widen the range of solutions available to meet the various environmental challenges. I believe that business has clearly demonstrated its commitment to finding these answers and has increasingly worked with other groups to do so.

In my view, the role of business is to offer people a better choice. In particular, this applies to environmental and social issues such as climate change and the immediate challenge of reducing greenhouse gas emissions to avoid the risk of global warming.

For the past five years we have argued that, for all the uncertainties of the science surrounding climate change, there is now a strong case for precautionary action. It is an undeniable truth that people link energy and pollution and also believe that companies should raise their heads and raise their aspirations. We know there are no simple answers and we realise that oil and gas companies cannot solve these issues on their own. What we can do is to make a contribution, and we can also help as part of a wider effort that includes all business.

Business is about targets and performance, and this has been our starting point at BP. In 1998 we set our own target of reducing emissions of carbon dioxide by 10 per cent from a 1990 baseline over the period to 2010 (allowing for normal growth this is equivalent to a nearly 40 per cent reduction) and challenged our people to come up with ways to achieve this. The response has been amazing. Today there are about 300 separate actions under way within the company and we are already halfway towards our target.

As well as setting a target we have also established an emissions trading system across all our 154 Business Units to ensure that the reductions can be delivered in the lowest cost way. In our view, trading is one of the most efficient methods of meeting the climate change challenge and our initial experiences have proved very positive. The system is not proprietary and we are happy to share our experiences with others.

There is a growing momentum behind the establishment of emissions trading systems, both nationally and internationally. We have been particularly pleased to see the excellent progress made by the CBI and ACBE in putting forward proposals for a UK domestic emissions trading system, and believe that, with the right level of government support for putting the system into practice, the UK can be an international leader in this area.

The second area where we think we can offer people a better choice – and at the same time help mitigate some of the perceived effects of climate change – is cleaner fuels. Air pollution is an immediate problem in scores of cities around the world. We believe we can make a contribution to improving this situation by making a new offer of fuels that are free of lead and fuel which contains ultra-low levels of sulphur. So far such fuels have been introduced in 59 cities worldwide and we hope to increase the number to around 90 by the end of 2001.

A third initiative is the development of a new form of fuel – solar power. At the moment solar power provides only a tiny fraction of the world's energy needs. But this will change as the market develops and technology advances. Perhaps it will take a decade or more for solar to become competitive with hydrocarbons as a means of generating power. But we think solar will be an industry of the 21st century and over the last five years have invested US $150 million in the business to the point today where we are one of the largest manufacturers of photovoltaic panels in the world. We have also begun to use solar to power our service stations in several countries, including the UK, and are now the world's largest consumer of solar power.

These are just three illustrations of what one company is doing to mitigate the potential effects of climate change. What they emphasise, in my view, is that the role of business now is to take the initiative – to recognise the challenge and to provide choices that help customers and governments avoid trade-offs between growth and environmental care which no one wants to make.

None of the initiatives we are taking represents instant solutions. But our experience so far is that if you take one step another becomes possible. As a great American once said: 'Making progress is simply about delivering something that people desire. . . but believe to be beyond reach.' That is exactly what we are aiming to do.

Sir John Browne was appointed Group Chief Executive of BP in 1995. He is a non-executive director of Goldman Sachs Group and Intel Corporation and a member of the supervisory board of DaimlerCrysler. He is also a director of Conservation International and a member of the Asia Pacific Council of the Nature Conservancy.

BP is one of the world's largest petroleum and petrochemical companies. Its main activities are exploration and production of crude oil and natural gas; oil refining; gas marketing and power production; retailing and transportation; and the production and marketing of petrochemicals. It also manufactures and sells photovoltaic panels for solar power generation. A truly global company, BP has well-established operations in Europe, North and South America, Asia, Australasia and Africa.

BP's policy goals are simply stated: excellent and ethical business performance; benefit to the wider community; no accidents in its operations; no harm to people; and no damage to the environment. To achieve these demanding goals BP sets measurable targets, submits the results to external verification and publishes reports on progress.

BP
Britannic House
1 Finsbury Circus
London EC2M 7BA

tel: 020 7496 4000
fax: 020 7496 4630
website: www.bp.com

7

Kyoto Targets: Their Impact on Business

Aidan Murphy,
Shell International, and
Dorje Mundle,
PricewaterhouseCoopers

What Kyoto means for business

At their most fundamental level, the emissions reductions targets contained within the Kyoto Protocol result in a price being placed on what was previously, in financial terms at least, cost free. They represent a new stream of costs that companies in Annex B countries face, which affects the economics of individual projects and wider corporate operations.[1] This raises two key questions: how large will these costs be, and how will they affect competitiveness?

At present the answers to these questions are not clear, creating an unquantifiable and continuing risk for business, although it is clear that the impacts will be greater for carbon-intensive industries. Not only are such businesses more affected by the amount of greenhouse gases (GHGs) that are associated with their operation, but they frequently operate by making very sizeable investments (capitalisation) in long-lived infrastructure (such as rigs, refineries, etc) that have prolonged payback periods. The uncertainty surrounding the future cost of carbon represents a potentially significant risk to these long-term investments. The appraisal of this risk is not entirely straightforward.

---

[1] Annex B to the UN Kyoto Protocol – generally, although not necessarily, OECD countries, in particular economies in transition, Russia and Ukraine. Sometimes referred to as Annex 1 countries.

Costs that are imposed on entire business sectors will to an extent be passed down the supply chain. The degree to which this occurs will depend on the competitive environment.

Managing the impacts

The first step in managing any risk is to understand it. This needs to be carried out at two levels: in the regulatory framework and at company level.

Regulatory framework

The cost of carbon will be determined by a combination of international, national and local policy and regulatory frameworks. At this early stage these frameworks, and hence the cost of carbon, are highly uncertain; the ratification of Kyoto is yet to be achieved, and the detailed rules and modalities contained therein are still in the process of being defined. Moreover, the policies and measures that individual nations adopt to meet their Kyoto targets are still emerging. Some countries are further ahead than others; Norway already has carbon taxes of 40–50 euro/tCO_2e, while in the UK there is a raft of measures such as the climate change levy and the UK emissions trading scheme. Furthermore, it is not possible to know what further targets will be agreed for subsequent commitment periods. However, it is clear that market-based policies will deliver very much lower compliance costs for both business and society in general. Ultimately this will be good for business, good for society and good for the environment.

Company level

Despite these uncertainties, it is nonetheless possible to make some progress in understanding the potential risk exposure at the company level. This involves monitoring and measuring emissions, ascertaining the marginal abatement costs (MACs) for CO_2 and other GHGs, which are currently poorly understood. Once companies are better informed on the range of MACs across their operations they are able to identify the most economically efficient abatement options, thus making better-informed investment decisions.

This can be a significant and time-consuming undertaking, and it may take many years for full MAC data to emerge for large and diverse corporations. Therefore policies and tools are required to speed this process and assist in reducing emissions in the meantime.

Reducing risk

One such tool is emissions trading, which helps reveal MACs while concurrently reducing emissions in a cost-effective manner. The Royal Dutch/ Shell Group introduced an emissions trading programme called STEPS (Shell Tradable Emission Permit System) in January 2000, which is proving a useful mechanism for revealing MACs while contributing to the Group's commitment to reducing GHG emissions to 10 per cent below 1990 levels by 2002.

However, while emissions trading helps business to position itself better to face future carbon constraints, it does not help it deal with this uncertainty and risk when making long-term investment decisions. To address this, Shell has taken steps to limit the risks that investments face, by applying a system of carbon values. This is designed to quantify risks under different scenarios, thus enabling systematic responses.

Carbon values represent the current and expected future financial costs from legislative constraints on the emission of GHGs and the financial benefits governments may provide for lower-emitting technologies. They are the costs that emitters of carbon are expected to have to pay to continue emitting; or they are the revenues expected to be available to companies that take action to reduce carbon emissions. Because the expectation is that carbon credits will be bought and sold, shadow carbon values also represent the current or expected future price of carbon in whatever market(s) it emerges.

Applying carbon values means modelling the potential impact of carbon prices on the economic performance of a business activity and identifying the financially optimal programme of measures to respond, as part of the normal business review process.

It addresses two questions:

- How much of a business risk or opportunity would carbon prices pose for the investment?
- How will the project effectively manage the risks and capture the opportunities? In other words:
  - How much room does the project have to lower its financial risk?
  - What early actions should it take to lower its cost of carbon abatement or capture revenue opportunities?
  - Is there a plan and the capability in place to implement these measures?

In addition, applying shadow carbon values means, over time, lowering the overall risk to the Group's portfolio and optimising competitive advantage from the expected emergence of carbon prices. Under this system Shell screens projects that exceed threshold investment and/or emissions sizes, and applies a range of values for the costs of carbon to the base case project economics of potential investments.

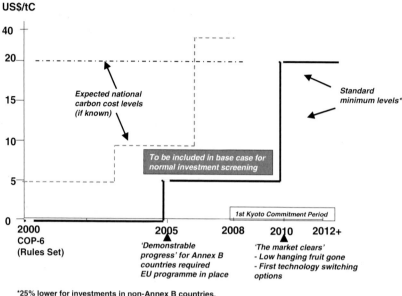

US$/tC

Figure 7.1 *Base case carbon cost levels for modelling*

Learning process

Adapting to this new framework requires raising internal awareness and organisational learning. This is a major task for Shell with 100,000 staff in 130 countries, but significant benefits can arise from this process, notably the identification of 'no regrets' and low-cost abatement options. This learning process and better understanding of MACs is a key enabler to achieving the dynamic efficiency that the use of economic instruments seeks to achieve. It complements standard, sound business management and reinforces the integration of sustainable development principles into Shell's core operations.

New opportunities

In a world of changing societal concerns and demands the private sector will need to respond by supplying new products and services, hence creating new business opportunities. Consequently Shell is taking strategic decisions, redefining its role away from a purely hydrocarbon-based business towards a wider portfolio of energy services. These opportunities can arise from the deliberate creation of new markets (eg the Kyoto market for CERs, emission reduction units (ERUs) and assigned amount units (AAUs)), and the changing balance of incentives and costs as markets grow and new technologies are developed and refined. Shell has recognised the potential growth of renewable energy in a carbon constrained future, and is in the third year of a US $500 million five-year investment plan to make a profitable business from renewable resources.

Conclusions

Kyoto presents an uncertain but potentially significant risk that will, if ratified, affect the competitiveness of some business sectors. It is essential that regulation moves further towards market-based approaches that minimise costs and increase environmental effectiveness. The Kyoto mechanisms are an important step in this direction, but it is critical to ensure they are designed with a full appreciation of how competitive markets and corporate investment strategies can be harnessed to deliver results consistent with the objectives of the Framework Convention on Climate Change (FCCC).

However, the risks presented by Kyoto can be significantly mitigated if steps are taken to understand how they will manifest themselves, and corresponding measures taken to reduce potential liabilities. Moreover, Kyoto represents the evolution of the prevailing economic conditions in which business operates, and this presents new commercial opportunities. These may be particularly important for the early movers that gain competitive advantage, but are predicted to be significant in the wider global economy.

Aidan Murphy heads the climate change programme for Shell International and is the primary spokesperson on such issues. Additionally, he is responsible for the delivery of the company emission trading system, project programme using the clean development mechanism (CDM), use of carbon values in investment screening, and development of a technology strategy for greenhouse gas emissions.

Group Climate Change Adviser
Shell Centre
London SE1 7NA

e-mail: webmaster@si.shell.com

The **Royal Dutch/Shell Group of Companies**' objectives are to engage efficiently, responsibly and profitably in the oil, gas, chemicals and other selected businesses and participate in the research for and development of other sources of energy. Shell companies are committed to contribute to sustainable development.

www.shell.com

Dorje Mundle formerly worked in Shell International's climate change team assisting with policy analysis and development, with a special emphasis on the Kyoto mechanisms.

He now works with PricewaterhouseCoopers where he specialises in advising on climate change, sustainability and environmental strategy.

PricewaterhouseCoopers
Global Environmental Services
Southwark Towers
32 London Bridge Street
London SE1 9SY
e-mail: dorje.mundle@uk.pwcglobal.com

PricewaterhouseCoopers' Global Environmental Services practice comprises 350 environmental management professionals located worldwide. GES works with business and government to manage the opportunities and risks presented by the environment and sustainable development, and helps define their strategic vision for environmental management and sustainability.

PricewaterhouseCoopers
Global Environmental Services
Southwark Towers
32 London Bridge Street
London SE1 9SY

tel: 020 7804 3000
fax: 020 7378 0647
website: www.pwcglobal.com

8

Allocating Permits for Emissions Trading

Bill Kyte,
Powergen

Introduction

In the past decade there has been increasing attention to the use of fiscal instruments in the environmental arena, and environmental economics is now a mainstream subject.

Trading is one of our oldest activities and basically only requires two elements – a tradable commodity and a willing seller/buyer relationship.

Emission trading schemes allow participants to buy and sell permits to emit pollutants, the overall number of permits having been fixed in advance. This enables environmental targets to be met in a more flexible manner by the participants, thus reducing costs.

There have been a number of different schemes in the past but the Acid Rain Program in the United States has been the largest and most successful in the world. The Program has achieved a strict environmental goal of reducing sulphur dioxide emissions and results have shown that compliance costs have been greatly reduced.

Under the Kyoto Protocol and the EU 'burden-sharing' agreement the UK will have to reduce its emissions of greenhouse gases by 12.5 per cent from their 1990 levels averaged over the period 2008–12. One of the mechanisms agreed within the Protocol is international emissions trading between countries, with binding targets.

In the UK business has been interested in exploring the potential of emissions trading and in October 1999 a CBI/ACBE Emissions Trading Group working with government officials presented outline proposals for a UK emissions trading scheme that, in principle, would be open to all companies operating in the UK which committed themselves to binding GHG limits agreed by government.

Allocation process

Allocation of permits is the critical issue in setting up an emissions trading scheme, since the allocation process can create significant commercial distortions and can have a marked impact on company valuations.

There are a number of possible ways to allocate permits, and although none is optimal for all players the allocation process needs to be as equitable as possible in order not to introduce undue distortions in the economy. It must also recognise current and future levels of business output and especially previous actions taken to reduce emissions of greenhouse gases.

Allocation principles

Since allocation will be a crucial issue involving significant negotiations between government and different sectors of industry, it is important to identify a set of principles about which agreement can be reached. It is just as important to identify areas that are likely to be contentious and which will require solution, probably either by pragmatic resolution through negotiation or by government decree.

It is important that there is absolute certainty through time and that any changes to the procedures should be defined from the outset, though a review mechanism should not be precluded.

The allocation methodologies in the next section are the building blocks upon which this process can take place.

Allocation methodologies

There are a number of methodologies for allocating initial and subsequent allowances, each with inherent advantages and disadvantages. Whatever the method chosen, for market confidence and business investment cycles, the mechanism of allocation should extend into the future as far as possible.

The purpose of this section is to describe the salient characteristics of the different allocation methodologies that could be used as the viable basis, or bases, for the allocation of emission allowances. Even a cursory analysis shows that there is no 'perfect' methodology and much will be determined by other factors, such as the government's overall economic, social, energy and industrial policies. Their advantages and disadvantages therefore accompany the descriptions of the methodologies in order to act as an agreed basis upon which an appropriate scheme, which takes account of pertinent factors, can be constructed.

Auction

According to economic theory the auctioning of permits is the simplest and most efficient allocation process, rewarding the efficient and penalising laggards. It is also attractive to governments that are able to extract the maximum rent from the process.

However auctioning does have significant practical disadvantages including the instantaneous raising of costs leading to market distortions and loss of competitiveness on a national and international scale. These distortions still occur even if the revenue is recycled.

Auctioning is regarded as taxation by another name and makes it extremely difficult for business to 'buy into' the process. It thus cannot be considered the primary methodology.

'Grandfathering'

The 'grandfathering' of allowances involves the gratis issue of emission allowances based on current, or historical, emissions or on extant regulatory requirements, thus giving explicit recognition to the implicit emission rights granted under a 'licence to operate'. The Kyoto Protocol was agreed on this basis but it is recognised that, in order to engage the developing nations, there will be a long-term need to move towards a non-historical basis.

As with the Kyoto Protocol, successful existing trading systems and most proposed systems have or are being based on a grandfather or modified grandfather concept in order to ensure a sufficient degree of acceptance and to avoid abrupt market distortions.

Grandfathering has a number of variants and can be flexibly modulated. The allocation can be based on a single year, an average of several years, or, for a future basis, on a 'rolling' average over a number of years. Basing the allocation on a historical average over a sufficient number of years would allow credit for past action. A minimum period of five years (or longer if certifiable data are available) is suggested. A 'rolling' average method could then be used post-2012.

The more complex variants significantly alleviate some of the perceived disadvantages of this method of allocation.

Negotiated agreements

Negotiated agreements between sectors, or individual companies, and government can form the basis of allocation. In reality, in the ultimate, all forms of allocation are based on some form of negotiated agreement in order to achieve business and government 'buy in'. The various allocation methodologies then form the basis of the negotiation process.

The introduction of the climate change levy and reduced rates for the levy for certain sectors in return for reaching agreements with government mean that negotiated agreements will form at least part of the UK framework for emissions trading. These must be accommodated into any wider scheme.

The major advantages of negotiated agreements lie in the buy-in by sector/ company/government and the potential flexibility that they offer to business. The major disadvantages are that reaching agreement is a long and difficult process and needs some form of contractual basis involving incentives/ penalties. There is also more potential for legal challenge than with other methodologies, as well as significant equity concerns. If the process is not completely transparent there may be concerns of commercial confidentiality.

Benchmarking

Benchmarking as an allocation methodology would appear to have the best intellectual credentials but, in practice, can be so complex that it can only be applied in the longer term due to the need for credible and transparent data. Benchmarking can be applied at many different levels and in some cases may need to be applied at the individual site level.

One of the major advantages of benchmarking is that it can, in principle, be applied to both credit and allowance trading. However it is extremely unlikely that credible verifiable data are available across all sectors and therefore benchmarking can only be regarded as a long-term option. It could form the basis post-2012.

Combination methodologies

Most successful systems have been broadly based on some form of grand-fathering, with appropriate modifications to ameliorate the disadvantages. These can take the form of reserving a small proportion of allowances for sale or auction in order to stimulate the market or to allow for new entrants, or take the form of a small permit levy on banked permits to discourage hoarding.

Recommended allocation process

The Emissions Trading Group recommended that the allocation up to 2012 be based upon a grandfather methodology with a small additional portion (1 per cent) provided by the government for auction, in order to stimulate the market and cater for new entrants. The government-brokered negotiated agreements have, by their nature, a built-in allocation.

It also recommended that the allocation, post-2012, should be based on benchmarking, 'rolling' grandfathering, auction or a combination of these methodologies.

Dr Bill Kyte is Head of Corporate Sustainable Development at Powergen and Chairman of the International Chamber of Commerce (ICC) UK Environment Committee.

Powergen supplies gas, electricity and telephone services to residential, business, corporate and government customers in the UK. It is a market leader in providing combined heat and power plant to industrial clients and trades electricity, gas and oil in UK and European markets.

Powergen plc
Westwood Way
Westwood Business Park
Coventry CV4 8LG

tel: 024 7642 4000
fax: 024 7642 5432
website: www.powergenplc.com

Section III

Financial, Legal and Insurance Issues

9

Linking Corporate Financial and Environmental Performance

David Edwards,
Andersen Consulting

Can businesses protect the environment and their revenues without one or the other suffering? It was in the early 1990s that business first started to take seriously the threats presented by the growing concern for the environment and the impact that business was having on the world around it. It was only in the late 1990s that the most proactive companies started to turn those threats into opportunities and create scenarios where both company and the environment benefit.

The motivation for companies to take the path towards sustainability may be encouraged by a variety of factors. Some are negative and reactive, such as the fear of non-compliance or the wish to avoid bad publicity; some are positive and proactive, such as new market opportunities and resource cost savings. The World Business Council on Sustainable Development, which includes major oil, mining and chemical companies among its 120 members in 35 countries, is testament to the willingness of modern companies to engage in dialogue with key stakeholders and play an active role in setting the environmental agenda for the future.

The reconciliation of the goals of profit maximisation for shareholders and protection of the environment is an issue that has been keenly debated for a number of years. On one side has been the view that environmental spending has little impact and that initial benefits give way to diminishing returns on environmental projects, with first-mover advantages being eroded as more and more companies adopt environmental programmes. Other commentators have insisted that systematic improvements in environmental performance will lead to improvements in the financial bottom line through a variety of factors such as efficient waste management, premiums on green products, improved public image and the avoidance of non-compliance with the growing scope of environmental legislation.

Empirical investigation in the US, where reporting requirements have provided the necessary comparable environmental information, has suggested that there is a clear and positive correlation between company environmental and financial performance. In 1995 researchers at the Investor Responsibility Research Center, Washington DC, compared return on assets, return on equity and total returns to normal shareholders for low and high pollution portfolios. They found that in more than 80 per cent of the portfolio comparisons the low pollution portfolio outperforms the high pollution portfolio. They concluded that it does not appear that investors who construct a balanced portfolio of good environmental performers will pay a penalty in terms of market performance. The researchers acknowledge however that it may be that firms are good environmental citizens because they can afford to be. There is another, more pragmatic way of looking at the benefits of a green approach, taken by those who advocate 'clean' technology in contrast to 'clean-up' technology. A firm is not in business to turn perfectly good and expensive inputs into polluting waste that at best is thrown away or at worst can increase disposal costs.

My own study, completed in 1996, was the first to examine quantitatively whether UK companies can profit from environmentally proactive strategies. The study quantitatively links environmental and financial performance in order to draw conclusions as to the bottom-line impacts of green company policy.

The financial results of 50 green companies, expertly assessed by Jupiter Asset Management as being environmentally the best in their sectors, were compared over a five-year period with those of non-green companies in the same sectors. The sectors studied were: building merchants and materials; electrical and electronic equipment; engineering; paper, packaging and printing; health care; food retailers and support services. The financial indicators used are return on capital employed and return on equity.

At the first stage of analysis the financial performance of each green company was compared to the average financial performance of a number of non-green companies of similar profile in the same sector between 1992 and 1996. At the second stage of analysis each green company was compared to the best financial performer from the non-green sample in the first stage. By selecting the best financial performer from the non-green sample, stage two provides a more rigorous examination of any results obtained at stage one.

The results from the study suggest that:

- there is a positive link between the environmental and financial performance of companies in all sectors investigated;
- over two-thirds of green companies perform better than their non-green counterparts in 1,200 direct comparisons of green and non-green companies at stage one of the analysis;
- the green companies perform as well as the non-green companies even at the more rigorous stage two of the analysis, where the best financial performers are selected from the non-green sample for comparison.

In most cases it appears that companies which pursue environmental improvements do financially better than their non-green competitors. At the very least this study shows that there need be no financial penalty for corporate environmental excellence.

There are good underlying reasons for a positive correlation between environmental and financial performance. More sophisticated environmental journalism and information exchange on the Internet are providing more transparency of pollution and environment liability issues. Banks and insurers are more willing to deal with companies that take the legislative threat seriously. Financial markets are more willing to invest in cleaner companies. The best and brightest people are more willing to work for a company with an unsullied environmental image. Green consumerism is maturing and the general public is using the consumer vote to influence corporate attitudes. Economic instruments, both incentives and penalties, are being used more and more to encourage constant improvement.

Any one of these factors might be considered unconvincing as a driver for environmental proactivity if viewed alone. They become indisputable when viewed as a whole and when their synergy is considered.

A trend has emerged in the last few years for businesses to actively engage in not just the environmental discussion but in the ethical and social arenas as well. Corporate social responsibility and corporate citizenship are terms increasingly used by enlightened businesses to encompass issues such as human rights, employee rights, environmental protection, community involvement and supplier relations. Andersen Consulting has worked with companies such as BP Amoco, SmithKline Beecham and Coca-Cola to develop a framework for the practical implementation of corporate social responsibility and corporate citizenship. This approach helps companies to get it right in their dealings with various interests impacting on the business, identifying the priorities and benefits – to the community as well as the company. This can be especially important for market entry strategy in emerging economies.

It is clear that companies will benefit relative to their competitors, from the ability to mitigate the risks and take advantage of the opportunities, provided by the move towards more sustainable economic development.

David Edwards read biology with management studies at Imperial College, London. He completed an MSc in environmental technology at the Imperial College Centre for Environmental Technology in 1996, specialising in business and the environment. He is author of *The Link Between Company Environmental and Financial Performance*, published by Earthscan. He currently works for Andersen Consulting in London.

Andersen Consulting
2 Arundel Street
London WC2R 3LT

tel: 020 7844 4000
mobile: +961 181786
e-mail: davideuk@aol.com

Andersen Consulting is a US $8.3 billion global management and technology consulting organisation whose mission is to help its clients create their future. The organisation works with clients from a wide range of industries to link their people, processes and technologies to their strategies. Andersen Consulting has more than 65,500 people in 48 countries. More details can be found on the Internet at www.ac.com.

10

Green Funds and Their Growth and Influence on Corporate Environmental Strategy

Charles Millar,
Jupiter Asset Management

Development of Socially Responsible Investment in the UK

In the decade and a half since Socially Responsible Investment (SRI) funds were first launched on to the retail investment market in the UK two important shifts have taken place in the way that they are operated. The earliest funds, widely and accurately known as ethical funds, generally operated on the basis of 'ethical exclusions'. These funds were 'negatively screened' in order to ensure that no investments were made in companies which operated in sectors that were the subject of ethical concerns, such as armaments manufacturing or the publication of pornography.

However around the start of the 1990s the first important shift took place, as a number of funds adopted a more progressive approach, by adding 'positive' screens to their investment selection procedures. The purpose of this refinement was to identify, and then invest in, companies that demonstrate *good* environmental or community oriented performance.

As the 1990s drew to a close the second wave of change continued to roll through the industry. Some SRI funds are now building upon the more proactive approach adopted by positively screened funds, by attempting to use their position as shareholders to encourage the companies in which they are investing to improve their environmental or social performance.

A new emphasis upon engagement

This process of active shareholding – or 'engagement' as it is now known – is held by many investors to be an attractive approach to SRI, as it offers a means by which they can express their principles through their financial assets without restricting their fund managers' freedom to act – unlike the more traditional positively and negatively screened SRI funds. In some ways the shift to greater engagement is an obvious step for SRI funds to take because improvements in social, environmental and ethical performance can frequently have a beneficial effect upon a company's bottom line. Furthermore, as professional fund managers often have good access to company management teams and familiarity with the issues that companies face, they should be particularly well placed to raise such issues.

Progress down this route has been hindered by the fact that few fund managers specifically include 'the environment' in their remit. Indeed only a very few have a specialist in-house environmental department – a resource that many see as being a prerequisite for effectively incorporating environmental issues into the decision-making process. In addition, those who have started to engage with companies control only a small fraction of the City's funds under management. As a result the impact of engagement by fund managers has been limited.

However things are changing – there has recently been a recognition by government that it is appropriate for investors to look beyond pure financial issues when making investment decisions. This has led to the recent amendment to the 1995 Pensions Act, which requires, by July 2000, that pension funds' Statement of Investment Principles cover the extent to which 'social, environmental and ethical' considerations are taken into account when making investment decisions. This new requirement has given SRI a substantial boost. More particularly it has given engagement-oriented SRI a boost. This is largely because pension fund trustees, with their eyes nervously fixed on their fiduciary duties, find an approach to SRI that does not entail any restrictions upon fund management flexibility an attractive option to adopt.

As a result of raising SRI's profile among institutional investors – and in particular among the massive local authority pension fund sector – it is hoped that the government may have set in train an increase in the growth rate of SRI funds. With the assets of retail SRI funds currently totalling only some £2.5 billion (barely enough to buy the smallest FTSE100 company) there is clearly room for growth. Crucially, such growth should increase SRI fund managers' ability to take larger stakes in companies and hence enhance their influence over companies with which they may attempt to engage.

Engagement in practice

So what form does engagement take? At Jupiter Asset Management the SRI funds only engage on issues where efforts to 'influence and encourage' management are expected to result in financial as well as environmental benefits. Identifying such issues requires a detailed understanding of the company in question. Hence decisions on where and when to engage are only made after our in-house SRI Research Unit has consulted with the SRI fund managers.

Once we have identified an issue we meet with the management of the company in question and suggest what we think needs to be addressed. The issues upon which we engage may be either specific or general. The former may be similar to Jupiter's current action to persuade garden centre operators to minimise their sales of peat products. The latter may involve encouraging the adoption of more formal environmental management systems, the development of more detailed environmental reports or the review of the priorities of a corporate environmental policy.

Although engagement is a gradual process, which often requires repeated meetings with management before results are evident, we believe that some of Jupiter's actions have already been successful. For example, one leading quoted travel operator has acknowledged our input into its efforts to prioritise actions in its environmental programme, while a small cap printing company has recognised our assistance in undertaking its first supplier assessment programme.

The future

At Jupiter, engagement is expected to become an increasingly important component of SRI. To that end we have begun to produce regular reports on the results of our engagement with companies. We anticipate that other SRI funds will adopt this approach, and when they do it will become possible to benchmark what promises to be a substantial new force for influencing improvements in corporate environmental strategies.

Charles Millar is Assistant Director of Jupiter Asset Management and Senior Researcher in the Jupiter Environmental Research Unit. Charles has worked at Jupiter Asset Management since 1994. He formerly worked for an environmental consultancy and a number of environmental non-governmental organisations. He is a member of the British Standards Institution's Environmental Certification Advisory Council and has an MSc in environmental technology from Imperial College, London.

Jupiter Asset Management Ltd
1 Grosvenor Place
London SW1X 7JJ

tel: 020 7314 4770
e-mail: cmillar@jupiter-group.co.uk
website: www.jupiteronline.co.uk

Jupiter International Group is a fund management company based in London, providing investment services to clients of all kinds worldwide. The company has roots traceable to the early 19th century but dates in its current form from 1985, since when it has built a reputation for investment management with an emphasis on performance and 'tailor-made' client services. Jupiter was also one of the first fund managers to offer clients Socially Responsible Investment products. The principal wholly owned subsidiary of Jupiter International Group is Jupiter Asset Management Ltd, which is responsible for the investment management of pension funds, investment trusts, unit trusts, charities and private clients.

11

Directors' and Officers' Personal Liability Under Environment Law

Paul Sheridan,
CMS Cameron McKenna

Introduction

This chapter considers the risk of prosecution faced by directors and senior officers and managers under environment laws. While the indications are that the regulatory authorities are disposed to prosecuting individuals where the evidence is available, such prosecutions are in fact very few in number. So, for instance, the Environment Agency states in its Enforcement Policy that where an offence results from a company's activities the company will be prosecuted and the Agency will also 'consider any part played in the offence by the officers of the Company, including Directors, Managers and the Company Secretary. Action may also be taken against such officers (as well as the Company)'. This action may include disqualification of directors. The Environment Agency has prosecuted several individuals and this has resulted in custodial or suspended sentences. However, these have generally involved very small organisations (ie 'one-man bands') and the proceedings have generally been in the magistrates court. It is difficult to draw any meaningful principles from these cases. Curiously, virtually all of these personal prosecutions arise out of waste management activities (and not, for instance, following water pollution or permitting offences).

Statutory provisions

The statutory language for personal liability of directors and officers under environment legislation is fairly well settled. Indeed similar provisions will be found in other areas of statutory law, including for instance the Health and Safety at Work etc Act 1974.

An example of the standard statutory wording is Section 157(1) of the Environmental Protection Act 1990. It states:

> Where an offence under any provision of this Act committed by a body corporate is proved to have been committed with the consent or connivance of, or to have been attributable to any neglect on the part of, any director, manager, secretary or other similar officer of the body corporate or a person who was purporting to act in any such capacity, he as well as the body corporate shall be guilty of that offence and shall be liable to be proceeded against and punished accordingly.

Similar provisions will be found in the Water Resources Act 1991 (Section 217), the Control of Pollution Act 1974 (primarily Scotland) (Section 87), Radioactive Substances Act 1993 (Section 36) and the Water Industry Act 1991 (Section 210).

There are a number of general points that can be made in relation to these statutory provisions:

- While the provisions are not expressly restricted to 'senior' management, in a 1992 judgment (*R egina versus Boal*) the court stated that this type of offence is aimed only at 'those who were in a position of real authority, the decision-makers within the company, who had both the power and responsibility to decide corporate policy and strategy'.
- There is no express distinction between executive and non-executive directors. As a matter of practice, however, non-executive directors are likely to be less exposed than other senior officers (see below).
- The prosecution must prove beyond reasonable doubt 'consent', 'connivance' or 'neglect' on the part of the director or officer.
- 'Consent' is taken to require knowledge of the matters in question together with an affirmative action or approval. It has been held that 'A man of course cannot be said to consent to what he does not know.' However, the courts have also indicated that shutting one's eyes to the obvious or allowing a person to do something in circumstances where a contravention is likely, without caring about the consequences, may be sufficient to amount to consent.
- 'Connivance' is less clearly understood but is taken to imply knowledge of and acquiescence in the offence committed. One judge said that to connive

is literally 'to wink at' or 'take no exception to'. Non-executive directors might be attacked under this provision.
- 'Neglect' is more readily understood, at least by judges. It has been said to imply 'failure to perform a duty which the person knows or ought to know'. In any event attempts at definitions of 'neglect' are probably best abandoned and a common-sense approach taken instead.

Delegation

In environment law the issue of how far a director or manager can properly delegate his duties to others in the management structure of the company is particularly ill-defined. Some help can be obtained from judicial decisions made under the health and safety legislation, but even there the issue remains uncertain. What is likely to be the case is that the law will accept delegation by a director of the exercise of certain of his duties to another provided that other director or manager has sufficient experience. The director who properly delegates matters will most probably be entitled to expect that the work he or she delegates is carried out in accordance with his or her instructions. It will not automatically amount to neglect if he or she fails to check the work is carried out correctly, albeit that where the circumstances demand a mere direction or warning by the director may have to be followed up to ensure that the necessary precautions or actions were in fact implemented. Obviously individual directors or officers identified in the company's environment policy (if any) as having specific responsibilities will be particularly vulnerable to prosecution if an accident occurs. Before accepting such responsibilities that individual should feel confident that he or she has the competencies required to carry out the duties together with management and financial support from the company to undertake those duties. The environment policy may also be used positively by that individual, as an instrument to make clear where his or her responsibility ends.

Insurance

Standard directors' and officers' insurance policies may be distinctly unhelpful. This is because they usually contain a pollution or environment exclusion, at least for gradual pollution (the terms of the exclusions vary considerably). There are specialistic policies available from environment impairment liability underwriters that may be valuable to those sectors where the risk is high.

Penalties

The penalties depend upon the level of the court in which the case is heard. In general, environment offences are 'triable either way', meaning that the case can be dealt with in a magistrates court or Crown Court. Both courts have the power to order the disqualification of a director. The following are the most common maximum penalties (some offences carry lesser and others carry greater penalties):

- **Magistrates court** – fine of up to £20,000 and/or six months imprisonment.
- **Crown Court** – unlimited fine and/or up to two years imprisonment.

Summary

Experience suggests that the environment regulators are not following a consistent policy with respect to the prosecution of individual directors and officers. The approach varies both geographically within the UK and between the different environment sectors. As mentioned above, to date most personal prosecutions by the Environment Agency have occurred within the waste industry. It is, however, possible to discern a trend towards a greater willingness to consider prosecuting directors and officers. In practice prosecutors make their own judgements about the gravity of the particular offence and the turpitude of the individuals. In order to identify any individuals at fault they will often ask to see the company's environment policy, manual, audits, management systems and training records (particularly if the industry is highly regulated). While at first blush any individuals named in such documents or with responsibility for follow-up actions may be exposed to prosecution, by the same token those documents may well assist particular individuals in avoiding prosecution. Such documents and underlying works, if well thought out, up to date, technically competent and well resourced could provide good evidence that the incident which occurred did not occur through individual fault, but rather was unforeseeable or a simple accident.

Paul Sheridan is a Partner in the Environment Law Group of CMS Cameron McKenna, which is a specialised group within the firm dealing with environment laws on a global basis but with a particular emphasis on the United Kingdom, the European Union and Central Europe. He holds an honours degree in law and is qualified as a solicitor both in England and in Queensland, Australia. He has been responsible for advising numerous UK (including government departments) and multinational clients on all aspects of environment laws in both contentious and non-contentious matters. Prior to practising in the field of environment law he specialised in insurance law and thus brings with him the risk assessment skills inherent in that field.

CMS Cameron McKenna
Mitre House
160 Aldersgate Street
London EC1A 4DD

tel: 020 7367 2186
fax: 020 7367 2000
e-mail: pfs@cmck.com

CMS Cameron McKenna is an award-winning, full service international commercial law firm advising businesses, financial institutions, governments and public sector bodies.

The firm has strong specialist expertise in areas such as finance and financial services; corporate; utilities and natural resources; real estate; environment; insurance and reinsurance; cross-border investment; technology, lifesciences and intellectual property; infrastructure and projects; human resources and pensions; competition and European law; arbitration and litigation. The expertise in these areas is reflected in the firm's listing in *Chambers & Partners Directory of the Legal Profession*, where it is ranked as a leading and highly regarded firm in 33 areas.

12

Environmental Risks and Opportunities: The Role of Insurance

Paul Pritchard,
Royal & SunAlliance

The increasing awareness of environmental risks has meant that there is significant interest in assessing whether insurance can, as in many other business areas, offer assistance to commercial organisations. The provision of cover for environmental risks nevertheless presents significant challenges for insurers (here examined in the context of the risks posed to the environment by the operation of commercial processes). The unhappy experience in the US environmental insurance market of the 1980s highlighted issues such as problems with legal interpretation of policy wordings and retrospective legislation, as well as identifying the need for environmental expertise in understanding the site and business specific nature of environmental risks. Building on improved risk assessment and management practices by both insurer and insured, as well as better clarity in policy wording, it now seems likely that we will be seeing increasing availability of desirable insurance products.

Historically insurance policies such as public liability cover made no specific reference to environmental risks; unanticipated pollution claims in the US led to a review of this position. From the early 1980s restrictions started to appear on policies in the UK for certain trades. The year 1991 saw the introduction by Association of British Insurers (ABI) members of wordings limiting cover to sudden and unintended events. Nevertheless, even with these restrictions, some cover in the UK is already likely to be in place via the public liability policy. A first step for any organisation contemplating environmental insurance will be to assess whether it considers this adequate or wishes to pursue additional cover, for example relating to own site clean up or pollution from causes other than sudden and unintended.

An ABI Joint Pollution Working Group comprising high level representatives across the insurance industry considered the coverage of sudden pollution risks in the UK with particular regard to public liability policies. A report was produced in 1998 *(Recommendations for the Underwriting of Pollution Risks)*, which was intended to underpin better quality underwriting practice. It suggests a tiered approach whereby a first stage questionnaire includes relatively simple questions on site use of materials, site surroundings and claims history. This can be followed up by more detailed queries as appropriate.

In considering specifically environmental insurance, contaminated land-related issues have often been the principal area of interest. Products have become available to cover aspects such as third-party clean up, cost capping on clean-up projects and encountering unanticipated contamination. The revision of the legal framework dealing with contaminated land has clearly raised the profile of this issue, and implementation of the Environment Act should also help to facilitate insurance cover by providing a clearer legal framework in which to operate. It is also likely that the debate will encourage a broader interest in the range of historical and ongoing environmental risks faced by commercial organisations.

Assurance that an organisation understands and manages its environmental risks is an important requisite for gaining insurance cover. In this sense the role of environmental management systems (EMS) such as ISO 14001 is of interest to insurers. There are however further considerations – namely that the insurer must be able to assess the level of risk and associated potential financial consequences. This is likely to require additional assessment, often on a site-specific basis given the geographical dependence of environmental risks. IT-based systems that can access publicly available data (eg on ground-water vulnerability or historical site activities) will help insurers in this process. Thus it can be seen that an EMS should facilitate insurance cover by ensuring adequate information is available for underwriting purposes and by providing ongoing risk management. Third-party certification (to ISO 14001 or EMAS) can give the additional comfort that systems have been independently assessed, however in most cases the complex and individual nature of the risks will mean that the insurer is likely to pursue his or her own investigations regardless.

The improved availability of risk management information from the organisation seeking cover, coupled with improved tools being applied by the insurers, means that some commonly voiced concerns may be addressed. It may not be necessary, for example, to commission further investigations at sites. This reduces the costs associated with upfront site visits and intrusive investigations prior to the offer of cover.

The magnitude and type of environmental risks faced by organisations vary greatly across sectors. The increasing flexibility demonstrated by insurers will therefore see products tailored to individual sectors and companies. For example Royal & SunAlliance has been investigating the specific contaminated land concerns of companies that purchase property for investment purposes. Their concerns can be very different from companies in the manufacturing sector.

In addition to 'pure' insurance products there will also be greater use of hybrid risk finance products. This might involve some mechanism to raise project funds as well as an insurance element, for example to cover cost overruns.

It might be asked whether insurance cover does in fact contribute to actual improvement of the environment (as well as meeting the commercial needs of insured and insurer). In all cases insurance involves another party (the insurer) with a stake in ensuring appropriate environmental risk control. Secondly the availability of insurance protection (against cost overruns for example) will provide balance sheet protection to enable the successful completion of projects that might otherwise be associated with excessive uncertainty, eg in remediation of a brownfield site rather than the choice of a greenfield location.

Dr Paul Pritchard is an environmental specialist within Royal & SunAlliance, working on corporate controls and supporting the underwriting of environmental risks. Following research on heavy metal contamination he concentrated on assessment of land and water pollution while working for the DTI. He joined the Royal & SunAlliance group in 1994 after experience with several environmental consultancies where his focus of attention moved towards environmental risk assessment and management.

Royal & SunAlliance
Leadenhall Court
1 Leadenhall Street
London EC3V 1PP

tel: 020 7283 9000
fax: 020 7337 5050
website: www.royalsunalliance.com

Royal & SunAlliance Insurance Group plc is one of the UK's leading insurers with premium income of approximately £10 billion and origins dating back to 1710. It operates in 55 countries worldwide, transacts business in over 130 countries and is in the top 30 of the FTSE100 companies quoted on the London Stock Exchange.

With innovative solutions and global representation, Royal & SunAlliance is equipped to service the insurance and related risk management needs of every kind of business, from the largest multinational corporation to small, local enterprises.

Section IV

Products, Producer Responsibility and Final Disposal

13

Integrated Product Policy

Brian Atkin,
Panasonic UK

Integrated Product Policy (IPP) and Integrated Pollution Prevention Control (IPPC): a warning

Before embarking upon the body of this article it must be pointed out that there must be no confusion between IPPC and IPP. Whereas Integrated Pollution Prevention Control focuses primarily on reducing the pollution caused by manufacturing products, Integrated Product Policy aims at limiting the environmental damage of a product throughout its life cycle, ie from its production to its scrapping.

Introduction

The European Commission is looking to change its (traditional) approach to environmental management, when applied to products, to a new approach based upon an IPP. This shift in paradigm is largely due to the fact that most of the design and manufacturing processes are either controlled, or regulated, by Standards and Regulations or Directives. This being the case, in order to continue to improve the level of environmental performance of products entering the European marketplace there needs to be further action focusing on the relationship between products and their burden on the environment throughout the whole life cycle of the product.

IPP and its appeal

There are a number of reasons for this growing interest in product-focused measures. From the environmental perspective, there has been a general realisation that the relative importance of consumption-related emissions and wastes is rising; a good example is that of the energy use of the electronics industry's products in the crucial 'use' phase of its life. Process-focused controls have been successful to a large extent in reducing industrial pollution, and remains a vital part of the strategy. Increasingly, however, there will be areas where the pay-off in terms of environmental benefit will be, proportionately, much greater from consumption-related rather than process-related initiatives.

The driver for IPP thinking is the big potential for capturing these benefits. In effect, it means looking at environmental policy in each product sector and deciding on the policy and market measures best suited to delivering the required improvements. This type of process will be inherently more complex than some of the traditional policy approaches that focus more on the control of sites and materials, because its application ranges across whole product sectors and is shaped to fit their particular market circumstances. As far as governments are concerned, however, this approach is attractive, despite its difficulties, because of the potential for high environmental gain.

A very short history to date

During the early part of 1998, the European Commission's DGXI (Environment Directorate) was the recipient of a report on Integrated Product Policy commissioned from Ernst & Young and the Science Policy Research Unit of the University of Sussex. Since then the Commission has held an IPP workshop – in Brussels in December 1998 – and an informal meeting of the Member State Environment Ministers – in Weimar in May 2000 – resulted in broad-based support for this policy initiative.

On the basis of the Brussels workshop, the Commission will develop a Green Paper in which it will outline proposals for European Union (EU) action. While it is true that the EU already has some product-related environmental policies (ecolabel, packaging waste with its focus on 'producer responsibility', the coming take-back legislation for electronics products and cars), these policies were developed without the foundation of a clear IPP concept.

What the industry thinks of IPP

Although the idea of an integrated product policy may sound radical it *could* be a better approach than that currently adopted by the Commission. Currently industry is faced with a burgeoning raft of environmental legislation emanating from different parts of the Commission. To those involved in dealing with the Commission first hand it appears that it has too many disparate and unco-ordinated agendas and this makes life increasingly more difficult, and costly, for those producing for the European marketplace.

Via UNICE (Union of Industrial and Employers' Confederations of Europe) European industry has suggested that, before any further policy initiatives are developed, the Commission should carry out a thorough review of the implementation, and effectiveness, of that environmental legislation which exists today. Only then should it consider developing something new. Industry is more than a little concerned that the Commission will not do this but will, instead, forge ahead with its development of IPP *as well as* continuing with all of its current pieces of environmental legislation.

What an IPP might mean for industry

For the future, and if this policy initiative is successful, it could well mean that specific elements of legislation/voluntary agreements, related to any industry's products, could fall under one piece of, yet to be devised, legislation. Typically this might bring together controls over materials usage, end-of-life management, energy consumption, etc.

If legislative action is taken, it will probably include measures to promote: (i) the integration of a product focus into all relevant EU policy (both environ-mental and general); (ii) the use of 'market mechanisms' to encourage better producer responsibility; (iii) the encouragement of markets for green products; and (iv) enhanced dissemination of environmental information. While it is too early at this stage to obtain a clear picture of the implications of IPP on industry, it is certain that the IPP approach could have important implications for the design, development, manufacturing, shipping, marketing and selling of products in Europe.

Brian Atkin has worked for Panasonic UK Ltd (PUK) since 1976. The first 17 years were spent within the service company, working initially as a service technician and later as a service manager at the UK headquarters. In 1993, in response to an initiative from a new managing director, he was appointed Environmental Affairs Manager. This provided a stimulating challenge as Brian's knowledge on environmental issues at that time was approximately zero! Since those early days he has gone on to be a representative of both his company, the UK and the European electronics industry, in trade association work aimed at addressing the environmental issues confronting the electronics industry today.

Panasonic UK Ltd
Panasonic House
Willoughby Road
Bracknell
Berkshire RG12 8FT

tel: 01344 853520 (direct line)
fax: 01344 853704
mobile: 07899 063085
e-mail: brian.atkin@mail.panasonic.co.uk
PUK website: www.panasonic.co.uk
MEI website: www.panasonic.co.jp

Panasonic is a brand name of Matsushita Electric Industrial Co Ltd (MEI) of Osaka, Japan. MEI is ranked number four in the Fortune 500 list of companies and had a consolidated global sales figure of US $63 billion for the year ended 31 March 1999. Worldwide 282,000 people are employed, with 14,200 of these based in Europe. MEI has 18 manufacturing companies, 20 sales companies and three R&D companies in Europe. Products manufactured cover the ever-widening spectrum of electrical and electronics products and include DVD players, televisions, communications products, laptop PCs, audio equipment, robotic insertion machines and electronic components.

14

Producer Responsibility: Is Getting Your Own Back a Sustainable Option?

Peter Jones,
Biffa Waste Services

A more efficient use of resources lies at the heart of the government's sustainability strategy. This might appear to be an obvious focus for the policy but many people – both in industry and society generally – are unaware of the relative scale of inefficiencies in resource conversion associated with developed Western economies. The reality is that 'Great Britain plc' consumes around 600–650 million tonnes of non-water resources each year (most in the form of carbon/energy and aggregates). Of that total around 10 per cent appear in the form of finished products purchased by consumers, and of that 60 million tonnes a mere 6 million tonnes is represented as private physical 'capital' six months after purchase. In effect we operate an economy where 99 per cent of resources end up as waste requiring neutralisation or storage elsewhere in the global environmental economy.

Clearly there are mitigating factors – for a start we are a service economy and much of our GNP output is thus weightless. No allowances are made for accumulations in the physical stock of social capital such as housing, infrastructure and so forth – there are no measurement systems in place. The net weight of the physical international balance of trade is also an arcane concept that defies measurement. Nevertheless the underpinning conclusion remains – the current equilibrium price for natural resources is very low when compared to the future anticipated equilibrium price (or cost) for neutralising the effects of squandering those resources in one or two generations' time.

The concept of producer responsibility – whereby manufacturers of physical products undertake liability for end-life waste management of those products – is a key step designed to bring those short and long-run equilibria into some form of proximity with each other. Producer responsibility cannot drive that process in isolation. It needs to form part of a holistic, integrated approach, creating a set of price signals that encourages improved resource efficiency on all players in society. Those other measures include end of pipe fiscal instruments (disposal taxes, discharge bans, etc), increased regulatory costs (integrated pollution prevention control, fines and prosecutions) and virgin input taxes on raw material consumption). The whole is held together by the 'glue' of traded permits at appropriate stages of the supply chain. Most of the latter tend to be demand-side issues – aimed at influencing consumption patterns for resources.

Producer responsibility – or Integrated Product Policy (IPP) – is far more interesting and significant in so far as it seeks to go to the heart of supply side factors by encouraging increasingly globalised and oligopolistic production sectors to undertake a fundamental review of the way their products are engineered, put together and marketed. IPP achieves this by the simple mechanism of obliging manufacturers of consumer capital goods – for instance – to incorporate end-life collection, dismantling and environmental neutralisation as another element of production costs.

The central challenge for industry is to recoup those costs in the selling price of its products without a threat to earnings. Currently those end-life costs tend to be funded by the last user in the chain – thus a transfer effect is called for whereby those costs are transferred to the first user in the chain in the selling price. In a price competitive framework the manufacturer thus has to focus on how the environmental cost of its product – as waste at some point in the future – can be minimised through redesign, light-weighting, reductions in toxicity and improved labelling of components (for later recovery and reuse).

Analysis of the cost of that transfer – as a percentage of turnover – suggests figures of around 2–3 per cent – lower for high added-value sectors such as pharmaceuticals, higher for areas such as newsprint or plastic packaging. Overall it may not sound much – but the cost of delivering environmental improvement through this route is potentially equivalent to existing levels of profit before interest and taxation (PBIT) returns in many industries.

Clearly something has to give – and the most promising area whereby this process can be kick-started lies in a growing acceptance by government that fiscal instruments which impact on demand side decisions might produce revenue flows that offset incremental supply side costs. Thus far landfill tax, the energy tax and road pricing scheme initiatives have all contained a limited reversion of funding flows aimed at developing best or better practice. Significant progress, however, is unlikely until *all* funding flows are remitted for sustainability objectives on an industry sectoral basis in a transparent framework. Remitting the majority of these funds in the form of lower national insurance contributions – which is what currently happens – lacks sufficient transparency and operates as a substantive transfer tax between capital and labour intensive sectors. In short it is in danger of acting as too blunt an instrument.

These issues may seem a long way from the business of emptying corporate dustbins – but our sector needs to anticipate the likely pattern of end process collection logistics and treatment technologies based on the impact of these processes across all industry sectors. The immediate impact lies in a shift of our customer base – from those at the end of the pipe to manufacturers and product producers. Already we are involved in dialogue with a number of key manufacturing companies which realise that the ability of the waste sector to manage waste material flows efficiently could be a major determinant of their future level of profitability. The nature of that exposure is a function of their product complexity, weight and level of hazard – and if you need help working out how that process might affect you in your business then developing a dialogue with your waste contractor might not be as daft as it sounds.

Peter Jones is Director of External Relations at Biffa Waste Services.

Biffa Waste Services is the leading waste business in the UK for the collection and disposal of industrial, commercial and domestic waste.

Biffa Waste Services Ltd
Coronation Road
Cressex
High Wycombe HP12 3TZ

tel: 01494 521 221
fax: 01494 463 368
website: www.biffa.co.uk

15

Extended Producer Responsibility: Electrical and Electronics Case

Peter Evans,
Sony UK

Background

In the early 1990s the European Union identified end-of-life electronics as a priority waste stream and since then has been attempting to introduce legislation to facilitate its recycling. In July 1999 it released the third draft of its proposed Waste Electrical and Electronic Equipment (WEEE) Directive. The Directive's major objective is to reduce the amount of waste electrical and electronic equipment arising, to reduce the amount of this waste going to landfill and to address its toxicity.

Within the third draft industry sees the following as key issues:

- **Scope** – All electrical products are covered, from industrial/commercial electronics to consumer electronics. It is far too broad and badly defined.
- **Ban of materials** – Various heavy metals and brominated flame-retardants are banned from being used in products.
- **Recovery and treatment** – Producers are responsible for setting up systems to provide for the recovery and treatment of WEEE.
- **Financing** – The costs for collection, treatment, recovery and environmentally sound disposal of WEEE from private households are to be borne by the producers.

The major issues for industry are:

- the historical waste issue;
- the financing issues;
- the material bans.

The issue of historical waste

It is this issue, more than any other, which has shaped the debate on the financing of WEEE to date. Industry has argued, and does still, that, as it was unaware of this 'problem' when it put products into the marketplace it made no financial provision for covering their end-of-life (EoL) phase. Also, there is a major issue with those market leaders in the industry today about having, potentially, to pay for the 'problems' of other, now defunct brands. It is this insistence upon being responsible for historic waste that has shaped the argument in favour of the consumer paying for the recycling of products at the time of purchasing a new product. The ideal situation for the consumer electronics industry would be to have a visible fee. When purchasing a new product the consumer would be charged a fee that would cover the cost of recycling EoL electronics. The fee would be passed from the dealer to a management organisation that would finance the recycling of end-of-life electronics. Consumers would be informed that they would be paying a fee for the recycling.

Compliance and financing issues

As can be seen from the above, producers are deemed responsible for all the costs associated with WEEE. The European Commission would like producers to be fully responsible for meeting all WEEE costs. The industry, however, favours a shared approach, with each actor in the supply and consumption chain taking a share of the costs.

Such an approach is already in existence in the UK as a means of discharging responsibilities under the Packaging Waste Regulations (PWR). This approach resulted in the introduction of the PRN (Packaging Recovery Notes), which were brought into being to prove that recycling obligations had been met and to provide financing for future investments in recycling infrastructure.

The government wanted the discharge of responsibilities under this piece of legislation done on a competitive basis and subject to market forces. In this respect a number of 'compliance schemes' came into being to address the anticipated demand. There are currently some 12 schemes in existence. Ever since its inception the PRN system has failed to do anything by way of building infrastructures; arguably, the reverse has happened. Industry is very concerned that the connections between a longer-term, strategic outlook and the free market are missing. The free market, generally, does not take the long-term view and will not achieve the long-term environmental objectives that the UK government is required to achieve.

For WEEE, Sony believes the introduction of competitive schemes (similar to PWR) may not give the lowest environmental burden or provide the long-term recycling requirements. Environmental experts from the electronics industry would suggest, in the strongest possible terms, that some of the aims of PWR, specifically those relating to the end-of-life phase, are always going to be in conflict with current competition requirements. This is primarily as a result of industry being encouraged by government to develop 'industry-led solutions'. Experience now coming out of some member states, coupled with proposals from others and from industry groups, tends to indicate that, when industry is asked to develop such a scheme it can invariably do so, and can present a proposal for a low-cost scheme. However, such proposals almost always fall foul of the relevant competition authorities.

We would suggest that there needs to be an open debate about this. It is becoming increasingly apparent that the existing way of doing things, ie on a competitive basis using market forces, might well not be able to deliver on the longer-term strategic environmental goals that the government needs to achieve. Sony believes a single scheme is the best option as it would standardise the level of recycling, ensuring continuous best practice.

By way of starting to look at ways in which the future WEEE Directive might be implemented, a number of major electronics manufacturers have been working on an industry-led solution – PRIMER (Producers' Institute for the Management of Electronics Recycling). PRIMER has produced a preliminary discussion document and proposal for product take back in the UK. It has been originated by key representatives of the electrical and electronics industry and is currently being distributed to electronics industry trade organisations for comment.

The material bans

Of particular concern to Sony is the impact the proposed lead ban will have on soldering practice. Current solder for electronics use typically has a lead content of about 40 per cent. Leaded solder is the standard for virtually the whole of the electronics industry, globally. Currently there are no commercially available drop-in replacements for the lead-based solder. Lead-free soldering has a number of disadvantages, in that:

- It requires higher operating temperatures and thus puts greater stress on components.

- It is not compatible with leaded solder.
- For larger assemblies the solder quality is not consistent.
- Cost is approximately double that of lead-based solder.

All manufacturers are believed to be investigating the replacement of lead-based solder, but it will be extremely difficult to introduce this into in *all* products by the Commission's deadline of 1 January 2004.

In conclusion

The major issue that Sony has to come to grips with is the one concerning funding of its responsibilities under the proposed WEEE Directive. Currently consumer electronics manufacturers do not generate sufficient profit levels to cover the WEEE costs. In this respect it is essential that the consumer pays towards the recycling of EoL electronics.

Peter Evans is the Senior Manager for Environment for Sony in the UK. He covers research and development, manufacturing, and sales and marketing environmental issues for the group. Peter represents Sony on the environmental committee of BREMA (British Radio and Electronic Equipment Manufacturers' Association) and BREMA on the environmental committee of the European Association of Consumer Electronics Manufacturers.

Peter is a chemical engineering graduate who joined Sony 25 years ago and has held many senior positions in engineering and manufacturing during that period.

tel: 01656 867536
fax: 01656 867592
e-mail: peter.evans@eu.sony.com

Sony United Kingdom Ltd is a subsidiary of the Sony Corporation of Japan. The sales organisation has been in the UK for 31 years, while Sony has been manufacturing in the UK for 25 years. The company has R&D, manufacturing, sales and marketing divisions within the UK. The manufacturing division is based in south Wales and employees over 4,000 people, producing TVs, computer monitors, set-top boxes and key components including cathode ray tubes for its other manufacturing plants in Europe. Sony UK employs a further 2,600 in R&D, sales and marketing, and general support services.

website: www.sony.co.uk

16

Packaging Waste Regulations: Do They Do Anything for the Environment?

Rana Pant,
Procter & Gamble

Introduction

The European Packaging and Packaging Waste Directive requires recovery of a minimum of 50 per cent, but not more than 65 per cent, of packaging waste by 2001, through recycling, energy recovery or biological treatment. A minimum of 25 per cent and maximum of 45 per cent of the total recovered packaging must be recycled. The following pages discuss whether the current European packaging legislation meets its stated objective of introducing a high level of environmental protection. It is pointed out that the current system imposes high costs on businesses and results in questionable environmental benefits.

Review of the status quo of packaging waste regulation

Procter & Gamble supports the twin objectives of the Packaging and Packaging Waste Directive (94/62/EC) regarding a high level of environmental protection and the functioning of the internal market. We agree that the Directive has pushed for recovery activities to be developed, particularly in areas where they were not taking place before. In many cases, however, it has prevented municipalities from finding the best solution based on local circumstances because the current packaging regulations have severe limitations with respect to both environmental and economic efficiency:

- There is a tendency to increase targets arbitrarily without appropriate research on economic, social and environmental consequences.
- Much emphasis has been placed on the approach of extended producer responsibility, which has led to a very high cost for recycling in many schemes.
- Packaging regulations are based on a rigid waste hierarchy instead of an integrated approach to waste management.

Maximisation of recycling or sustainable packaging waste policy?

Current EU waste policy tends to maximise recycling, whereas the aim should be to minimise the environmental burdens associated with packaging at a reasonable cost. Recycling is certainly one of the key tools that will achieve this aim, but it remains a tool. The results of recycling efforts have to be assessed against this aim (compare Gabola, 1999). A sustainable waste management system will have to achieve a balance between environmental effectiveness, economical efficiency and social acceptability.

A recent study attempted to evaluate what level of recycling of plastics packaging waste would be the most eco-efficient (APME, 2000). The study has not yet been peer reviewed, and therefore the results may be considered preliminary, but they highlight that a further increase in recycling of plastics packaging over the current rate of 15 per cent, to 50 per cent for example, will lead to significantly increased costs, in this case by a factor of 3, without achieving significant environmental benefits. This example highlights that there is an optimum level of recycling and that it is necessary to evaluate all possible environmental burdens before implementing recycling targets.

Extended producer responsibility versus shared responsibility

If manufacturers become responsible for the collection, sorting and recovery or disposal of their products there will be a tendency towards parallel or segregated waste management systems. Segregated waste systems lose the benefits of economies of scale, and synergy between different treatment options, so tend to be less efficient, both economically and environmentally.

Costs of extended producer responsibility systems

The absolute recovery fees for plastic bottles demonstrate the wide range that can occur for the same package execution in different countries (see Figure 16.1). The French scheme operates with fees that are lower by a factor of 30 to 40 than the German and the Austrian schemes and the Dutch and Dannish schemes operate without any fee.

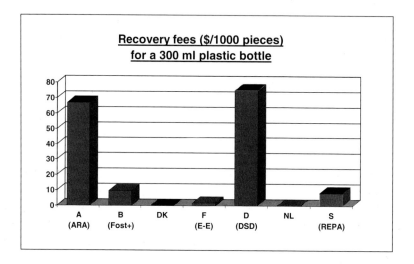

Figure 16.1 *Differences in recovery fees*

Source: Draeger (1997).

In contrast to extended producer responsibility, shared responsibility means that the owner of the product at each stage in the life cycle is responsible for any wastes produced that he or she directly controls, ie the disposer pays. This is key for giving the right incentive to divert material from final disposal. When a product passes from one owner to another in the life cycle the responsibility for any wastes subsequently generated passes with it. This shared approach allows responsibility for waste collection and sorting to be kept by a single entity (municipalities) so is compatible with an integrated approach to waste management.

Waste hierarchy versus integrated waste management

Decisions on waste management strategies have often been based on 'the waste management hierarchy' (waste reduction, reuse, recycling, composting, biogasification, incineration with energy recovery, incineration without energy recovery and landfill). Taken as a rigid hierarchy (rather than as a set of guidelines or a menu of options), it does not allow for the flexibility required when selecting the most environmentally effective and economically efficient method of waste management for a specific geography/location:

- There is little scientific or technical basis for listing the waste treatment options in this order. There is no scientific reason, for example, why materials recycling should always be preferred to energy recovery.
- The hierarchy approach is unable to compare the environmental advantages and disadvantages of the combination of different municipal solid waste treatment options.
- The hierarchy does not address costs and therefore does not lead to or promote economically efficient waste management systems.

Integrated waste management systems address the whole municipal waste stream, the materials to be recovered and the optimal treatment methods to be employed.

Figure 16.2 visualises that material recycling, biological treatment such as composting or biogasification and energy from waste are all potentially feasible. None should be ruled out. It must be noted though that integrated solid waste management systems ought to be developed at the local level because only then can they be optimised to the local conditions that differ from region to region and from urban to rural areas.

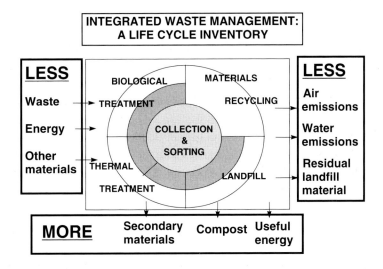

Figure 16.2 *Integrated waste management*

Rather than rely on the waste hierarchy, the environmental management tool of life cycle inventory can be used to find/develop the optimum solution for a specific location (McDougall and White, 1998; McDougall *et al*, 2000).

Requirements of a future-oriented packaging waste policy

The future formulation of packaging waste policy should be based on the criteria of sustainable development. That means:

- Packaging waste policy should be part of an overall waste policy. Waste policy needs to be aligned with overall environmental policy objectives.
- It must make a contribution to the three pillars of sustainability: environmental, economic and social considerations.
- It must define clear ecological goals instead of prescribing the means.
- It has to aim to achieve ecological improvement with the financial burden to the public and companies minimised.
- It must be open to changes in ecological and economic conditions and have a correspondingly flexible formulation.

References

Association of Plastics Manufacturer in Europe (2000) (APME) 'Assessing the eco-efficiency of plastics packaging waste recovery', summary report, Brussels.

Draeger, K (1997) 'Managing for producer responsibility', p 5 in *Take it Back! '97, Beyond Compliance: Global Issues for the next Millennium*, 17–18 November, Raymond Communications Inc, Alexandria, VA

Gabola, S (1999) 'Towards a sustainable basis for the EU Packaging and Packaging Waste Directive', speech given at the European Recovery and Recycling Association Symposium, November

McDougall, F R and White, P R (1998) *The Use of Life Cycle Inventory to Optimise Integrated Solid Waste Management Systems: A Review of Case Studies. Systems Engineering Models for Waste Management*, Gothenburg.

McDougall, F R, White, P R, Franke, M, Hindle, P (2000) *Integrated Solid Waste Management: A life cycle inventory*, Blackwell Science, Oxford

Dr Rana Pant joined Procter & Gamble in 1999 after obtaining his PhD in the field of life cycle assessment at the Darmstadt University of Technology, Germany. He studied environmental engineering with focus on waste management at the Technical University of Berlin, Germany. As a member of P&G's Global Integrated Solid Waste Management Team Rana deals with solid waste related packaging issues and integrated approaches towards waste management.

Procter & Gamble Technical Centres Ltd.
(PO Box Forest Hall No 2)
Whitley Road, Longbenton
Newcastle upon Tyne NE12 9TS

tel: 0044 (0)191 279 2656
fax: 0044 (0)191 279 2871
e-mail: pant.r@pg.com

Procter & Gamble (P&G) is one of the leading consumer goods companies worldwide, marketing approximately 300 brands to nearly 5 billion consumers in over 140 countries. These brands include Bold®, Ariel®, Crest®, Fairy®, Pantene Pro-V®, Always®, Whisper®, Pringles®, Pampers®, Oil of Olay®, Iams® and Vicks®. Based in Cincinnati, Ohio, P&G has on-the-ground operations in over 70 countries and employs approximately 110,000 people worldwide.

Section V
Mobility, Transport, Communication and IT

17

How Business can Respond to Green Transport Policy

John Elliot,
Pfizer

Travelling in our overcrowded island by road is becoming progressively more difficult, both in terms of increasing travel times and less predictability. This is especially true near our major cities. Most of us are also aware that motor cars produce significant amounts of pollution and contribute to global warming from so-called 'clean' carbon dioxide emissions.

Up to 1994 the traditional method of 'solving' traffic problems had been to widen roads, build a separate 'motorway' network and provide bypasses for towns. While the road programme has allowed many of us a lifestyle that we now enjoy, some road schemes were very unpopular because of the damage caused to the environment – whether in open countryside or in towns. Further significant expansions to the road network by any government would be extremely difficult.

Furthermore it was demonstrated, from work carried out in various countries before 1994, that by building more roads (especially in or near major urban areas) extra traffic was created. Thus transport can, in certain circumstances, be made worse for both motorists and public transport users! While this may seem difficult to grasp in the first instance evidence shows that, when a new road is built, people are willing to travel further to reach work, they give up public transport altogether to drive instead, and we end up in a vicious downward spiral in the transport arena. Put simply, more roads mean more car traffic, more people travelling by car means fewer on public transport, fewer people on public transport leads to higher fares and reduced frequency of service, which in turn means more people drive, thus adding to the congestion.

It is clearly the objective of this government (as of the last, post-1994) to try to establish a background where we can all benefit from a greener lifestyle. For those unwilling to make the shift it is necessary to provide some 'encouragement'. For expanding companies this encouragement also appears in the limited number of parking spaces being allowed with new developments and often a duty to produce green travel plans. As a business we realise that congestion and unpredictability of journeys is not only inefficient but also the stress on our staff causes frustration and lowers productivity.

Turning to a greener lifestyle in our management practices makes sound business sense in the longer term for the world, 'United Kingdom plc' and for individual companies. The benefits can also include being able to use more of our sites usefully rather than just for parking metal boxes. The individual travelling by train or bus rather than driving can use the time for working, reading or relaxing. Those walking or cycling can be fitter and healthier.

Pfizer is in the health care business and also has a set of well-instilled company values covering a range of areas, which embody our work and are embraced in our decision making. Care for the community is a prime example – ensuring that we recognise the area in which we are situated, where our rapidly growing site makes a green travel strategy particularly important to our current and future existence. This philosophy embodies the interests of our own staff, the general public and other organisations in east Kent – if not the whole of the UK.

Pfizer's transport strategy is tackling the delivery of 'green' travel by providing mechanisms to encourage all staff to move up in the transport hierarchy of:

- walking;
- public transport;
- cycling;
- carshare;
- motorcycling;
- single occupancy vehicle.

The strategy to deliver this has seven main elements, which are as follows.

Parking cash-out system

This is a system where everybody who arrives at work without a car will be rewarded to the tune of £2 per day. We are putting a substantial investment into this measure and believe it will have a lasting impact on travel behaviour.

Public transport improvements

We provide a free bus service for staff who live in the neighbouring town of Sandwich or who wish to travel there during the day. Within a few months of the service starting it was well used and popular. A 10-minute interval between services gives the sort of facility that encourages car drivers to leave their vehicles at home and use other modes of transport.

We have traditionally provided a works' bus system and this has been expanded significantly. In partnership with the local public transport operator, Stagecoach, we are now offering more destinations, longer hours of operation and a combination of improved service information, maintenance, cleanliness and comfort on the buses. As part of a package of initiatives with the operator we have integrated fares so that other timetabled Stagecoach services can be linked with our own contract works' buses for small add-on fares, and are also funding significant improvements to local bus services, not only for our own staff but also for residents in the local east Kent community.

Cycles and motorcycles

We are fortunate to be at the junction of two national cycle routes, coming from Dover via Deal to Sandwich and Canterbury, with an additional spur from Sandwich to Thanet. We have been working with the local authorities and Sustrans (the National Cycling Group) to encourage the best solution to the local cycle routes. We are also providing facilities such as showers, changing rooms, lockers and cycle sheds that are close to people's places of work on our large site. Motorcyclists also benefit from the on-site facilities and from the parking cash-out system.

Car share

Around 19 per cent of our staff currently share their car journey with colleagues, but we wish to increase this even further and are setting up a sophisticated computer system that will enable people to find travel partners. This is particularly important in the villages within east Kent, where we cannot provide bus services.

Infrastructure improvements

While we recognise that we are at the end of rail services linking the county to London, and immediate improvements to such services are unlikely to materialise, we are still campaigning. At our current rate of growth and location our demand for vehicular travel cannot be met on the very limited road network around our site. The local authorities recognise this issue and have planned improvements within the integrated transport policy to the local road network. We would stress that we are not asking for motorways but just a local road network to meet what is basically a new town employment zone, away from the main road network. We will also be encouraging any extra capacity to be reserved for buses and high occupancy vehicles.

On-site networks

Within our site, as part of our master planning process, we provide additional internal bus routes, cycle routes and walking routes that will be more attractive and treated above car travel to reinforce our 'green' messages.

General enablers

An important part of any transport or green travel plan is to ensure that everything works together and attitudes are changed. For example, at Pfizer we have an opt-out system for company cars and many of our staff enjoy an equivalent financial sum if they decide not to take up the option of having a company car. Furthermore we are making staff aware of the cost of motoring and are embarking on an extensive publicity campaign within the organisation, and east Kent, to explain all the reasons and advantages of our travel plans. Leadership in the last year from our two most senior directors who, cycle or use trains to work, has also been particularly helpful.

Pfizer has seen the merits of going forward with the green agenda. We recognise that culture changes take time and require considerable effort.

Community leaders in business and in the public sector need to embrace the necessary approach to travel, live it in their own organisations and explain it to their workforce. Tangible benefits for those adopting greener lifestyles need to be provided to employees, the general public, and from government or local government to organisations that are adopting a greener approach. Last but not least, anybody in a position of influence needs to ensure that government, local government and company policies work in a common direction.

Pfizer is a research-based, global pharmaceutical company with 5,500 employees in the UK.

Pfizer Ltd
Ramsgate Road
Sandwich
CT13 9NJ

tel: 01304 616 161
fax: 01304 656 221
website: www.pfizer.co.uk

18

Tesco's Integrated Logistics Management Systems: Creating Value Through Better Performance

Lucy Neville-Rolfe,
Tesco

Tesco is committed to serving the needs of the shopper. Its core purpose is continually to increase value for its customers to earn their lifetime loyalty. I believe that this simple business objective is the foundation of our success as a leading international retailer. As a company we are conscious that the concept of 'value' is far broader than competitive pricing and excellent range. Customers are increasingly demanding in what they expect of retail providers. Our environmental programme is therefore integral to our whole corporate philosophy of improving value.

Supermarket shopping, by its nature, reduces road congestion and fuel consumption. Large, fully stocked stores close to where customers live subtract from car mileage and add efficiencies of scale to bulk transport. Further gains are ensured by positive cost-reducing environmental programmes. From improving efficiency in the supply chain to upgrading refrigeration systems in stores, depots and lorries we pass on economic and environmental gains to the shopper. These initiatives alone have led to a reduction in our CO_2 emissions by 47,000 tonnes per annum.

For Tesco such 'win-win' scenarios are pursued not only for their environmental merit but also for sound commercial reasons.

Better logistics

Our Integrated Logistics Management (ILM) programme is our most recent success story in driving down costs and improving environmental performance. It has been widely praised: ILM was commended in the Energy Technology Support Unit in Best Practice 364 and was highlighted in the Department of Environment, Transport and the Regions' strategy document, 'Sustainable Distribution'.

The broad principle is very simple: to take empty trailer units out of the supply chain. Empty lorries cause congestion, consume energy and lose money. ILM seeks to remove them from the system, particularly in the return journey from store to depot.

Traditionally, Tesco used a system of 'primary' distribution to move goods from supply bases to Regional Distribution Centres (RDCs), using suppliers' or third-party vehicles. 'Secondary' distribution, generally undertaken by the Tesco fleet, delivered goods from the RDCs to stores. Both sets of vehicles normally returned empty.

To improve this situation, we introduced *'supplier collection'* and *'onward supply'*. Both of these initiatives entail the full integration of primary and secondary distribution functions. Supplier collection takes out the wasted journey when secondary distribution vehicles return from stores; now, Tesco vehicles convey a load from RDC to store, then travel on to a supplier to collect a primary load before returning to the RDC. Under 'onward supply', vehicles used for primary distribution from supplier to RDC are reloaded at the RDC to convey goods to a designated store. The general principle is illustrated in the figure below.

Of course, not all journeys can be operated within the parameters of these schemes; much depends on the relative location of stores, suppliers and RDCs. Nonetheless the schemes have had a major impact. We have cut 3 million vehicle miles, increased vehicle utilisation by 20 per cent and reduced CO_2 emissions by 4,600 tonnes. Tesco lorries now travel 85 per cent full on their journeys, an important international benchmark.

A further improvement is the introduction of independent haulage contractors to combine deliveries from several suppliers rather than each supplier having to deliver individually. This innovation has the potential to reduce the number of large goods vehicles required in the distribution chain by 25 per cent and to reduce empty running by 15 per cent.

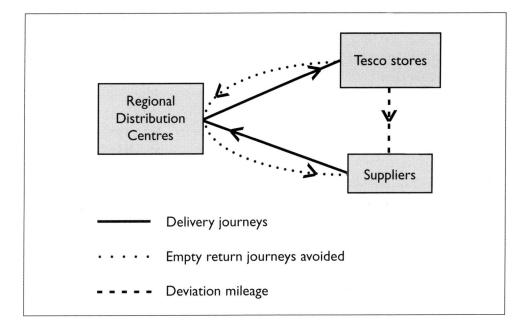

Figure 18.1 *'Supplier collection' and 'onward supply'*

Better design

New packing technologies and trailer configurations have amply returned significant investment. Traditional primary and secondary distribution functions required different vehicle designs. New vehicles have been developed so that those designed to carry pallets (used in primary distribution) may also carry store cages (used in secondary distribution) and so that temperature-controlled trailers may operate in both the single and multi-compartment configurations. The latter is important for secondary distribution, where different products need to be conveyed in relatively small quantities at different temperatures.

Recent research by Tesco has allowed the introduction of cages that travel throughout the supply chain from supplier to the aisle in the store and back again to the supplier. An in-house team has perfected the design of dolly wheels and cages so that units can be stacked by suppliers with produce that is then left untouched until the shopper picks it from the shelf. Produce is kept in pristine condition and time is saved at every stage of the supply chain.

Better emissions standards

It is not good business for our customers to associate Tesco with intrusive and polluting juggernauts. We have therefore developed state-of-the-art vehicles with our suppliers that use a new exhaust which traps pollutants and cleans emissions.

Better government?

Efficient distribution systems are critical to the success of our business. So, while we have demonstrated our commitment to responsible and sustainable corporate practice, we need also to ensure that environmental initiatives do not undermine our competitiveness and thereby increase prices. There is nothing so unsustainable as a business hampered by expensive regulation and poor infrastructure. In light of this we urge industry to press the government to act on three fronts:

- **Road charging and fuel taxes** – Through vehicle excise and fuel duties, commercial road use is already taxed at a very high rate – undermining the competitiveness of British business on the continent. We should continue to put pressure on the government to revise the fuel price escalator. We should examine very carefully any proposals for road charging, particularly where they involve differential rates for heavier vehicles. By discriminating against efficient and clean domestic hauliers the government is doing both the producer and the consumer a disservice.
- **Investment in transport** – Industry needs to press for more government investment in transport, vital for economic prosperity. More of the money that road users pay in taxes – some £31 billion per annum according to the Automobile Association – should be reinvested in new roads and public transport, maintained to a higher standard. Reductions in congestion would inevitably lead to lower emission levels.
- **Car-parking charges** – Industry needs to work hard to convince government that a tax on car parks would have serious repercussions. Such a tax, whether for customers or employees, would lead to people parking illegally or in residential areas to avoid paying a tariff in previously free car parks. Retailers would also be forced to increase prices in order to pay for car-parking to remain 'free' – pushing up shop prices while making no contribution to traffic reduction.

Better business

Through the implementation of 'win-win' environmental programmes Tesco has been able to make large efficiency savings. This can only add value to our service, on the one hand passing on cost savings to the customer and on the other ensuring that the shopper is buying from a clean and responsible business. Further initiatives and innovations will improve upon this impressive record. The right sort of thinking by central government could produce environmental and cost gains that at present are denied both to industry and the consumer.

Lucy Neville-Rolfe joined Tesco as Director of Corporate Affairs in 1997 after a career in central government. She is a member of the CBI's Economic Affairs and Europe Committees.

Tesco Stores Ltd
Tesco House
Delamare Road
Cheshunt
Hertfordshire EN8 9SL

Tesco is the largest retailer in the UK, with sales of £18.5 billion in 1998/99, employing 210,000 people worldwide. It operates 646 stores in the UK and 184 in Ireland, Hungary, the Czech Republic, Slovakia, Thailand, South Korea and Taiwan. It is a public limited company.

website: www.tesco.co.uk

19

Telecommunications Strategies for Future Growth and Flexible Working

Ian Wood,
British Telecommunications

Flexible working is not new, but its use is growing rapidly. Information and communications technology is enabling this growth, but the key drivers are economic, social and environmental. Fundamentally, flexible working enables people to be free from the constraints of location because work is sent to them rather than their having to travel to get it.

The cost to the individual and company of using cars is escalating through macroeconomic changes targeted to address local and global environmental problems, for example:

- fuel tax accelerator;
- changes to the company car tax regime;
- proposed congestion charging;
- planned toll roads;
- proposed taxes on work place parking.

The desire to mitigate the impacts of global warming and climate change on economic growth and social development is the biggest environmental driver. Cutting the growth in commuting, which wastes time, money and finite resources, will significantly reduce emissions that are contributing to global warming.

The changing shape of the family, and especially the trend for more parents to combine work with parenthood, is generating pressure for all employees to have more flexible options about the time they spend at home.

Companies competing to retain and recruit scarce skilled workers are having to rethink their business models to compete with flexible work styles on offer in the market.

Fundamentally flexible working is about a new business model (see Figure 19.1). Information and communications technology (ICT) is the enabler, but flexible working is really about releasing the potential of people, supported by technology, and developing a work style that is most effective and efficient for them.

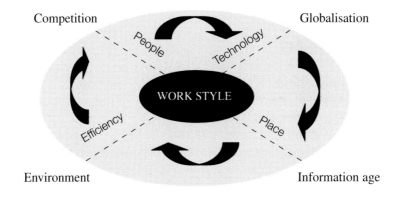

Figure 19.1 *Work style and a new business model*

It's about having the freedom to work anywhere (at any time), with access to all the resources associated with traditional offices. With this freedom work is not a place you go to; it's something you do. Travel is not eradicated by any stretch of the imagination. Social interaction is a critical part of work and travelling to meet people will always be part of most people's work style. But now it is no longer necessary to travel to access information or receive your telephone calls. Flexible workers typically divide their days between social days and quiet days – the latter working at home away from the interruptions of the office.

Flexible working brings a broad range of benefits, to businesses, individuals and society as a whole.

Benefits to the organisation

- saving on office accommodation and other employment overheads;
- increased productivity;
- reduced absenteeism;
- flexibility to respond to new competitive challenges quickly (and keep sales/ support people close to their customers);
- skills retention and a wider pool of potential employees;
- reduction in environmental impacts and costs.

Benefits to the individual

- less stress;
- reduced travelling;
- improved personal productivity/satisfaction;
- greater freedom to balance work and home demands.

Benefits to society

- reduction in pollution;
- reduced investment in transportation infrastructure;
- improved air quality;
- less need for migration of people to areas of better employment prospects;
- improved opportunities for part-time workers and carers.

Want to know more? Visit:

- **BT's Environmental Performance Report** – www.bt.com/epr2000
- **Working from Home** – http://www.wfh.co.uk/
- **Telecottages Association** – http://www.tca.org.uk/
- **Telework 2000 Conference** – http://www.telework2000.com/

Case study: BT Options 2000

BT presently has 4,000 workers based at home full time. This is a significant number, given that in 1993 the corresponding figure was just 100. There are over 40,000, BT people with the ability to remotely access e-mail and the company's intranet, and who work from home at least once a week. Tele-working was originally introduced as a result of an internal initiative, with the aim of reducing operating costs.

Options 2000 is the most recent flexible working initiative, launched in 1999, and mainly driven by the need to consolidate office space. However as the name suggests it is giving people alternative options, and the presumption is that if an employee wants to work flexibly he or she should be enabled to do so. The company plans to have 7,500 people home based by March 2002 as a result of the Options 2000 programme.

The initiative benefits from existing ICT infrastructure. The emphasis is on providing laptops, perceived as the most suitable option as they can also facilitate mobile working and working from customers' premises. Also in use are integrated fax, scanners and printers in one unit, ISDN lines and mobile phones. Audio and videoconferencing are also extensively used and reduce the need for travel.

Probably the most important aspect is the availability of the company's intranet, which is the largest in Europe. All processes and information in BT is managed through the intranet. For example product information, pricing schedules, project documentation, business expenses, pay slips, overtime payments, training, internal and external recruitment are just some of the things that are delivered only via the intranet. Remote access to the BT intranet network is guarded by encrypted passwords, which change every minute, and, if support is needed, BT employees have 24 × 365 help desk support.

Benefits of the initiative

It is estimated that work productivity has been increased by 20 per cent. Furthermore there is a positive impact on the environment, quality of work life, and employees' morale. It is also reported that managers who themselves are flexible workers regularly become even better managers, since their communication effectiveness improves even more.

Thanks to flexible working the speed at which BT's products reach the market is enhanced. Managers get their teams, required to develop new products, working together much more speedily by using flexible working.

A study demonstrates that BT home workers have reduced home-to-office miles travelled by car by an average 3,149 miles per annum, over 12 million miles per annum when extrapolated.

Impact(s) of the initiative

The biggest impact of the flexible working initiative in BT has been changing the ethos within the organisation from supporting buildings to supporting people.

The initiative has brought the need for changes for managers too. They need to work in a different way, and learn to manage people remotely. Of course managers still bring their teams together regularly to supplement virtual interaction with a physical one. In addition the practice widely adopted in BT is to manage by objectives, which can be effectively applied to flexible working.

Ian Wood works in the BT Environment Unit and is responsible for developing the environmental benefits' propositions associated with BT's products and services. Before joining the Unit Ian worked in supply chain management with roles varying from materials management to account management.

BT Homeworkers
PP: HW D182
PO Box 200
London N18 1ZF

e-mail: ian.t.wood@bt.com
tel: 01977 591660
fax: 01977 591661

British Telecommunications is one of the world's leading providers of telecommunications services and one of the largest private sector companies in Europe. Its principal activities include local, long-distance and international telecommunications services, mobile communications, Internet services and IT solutions. In the UK, BT serves 28 million exchange lines and more than 7 million mobile customers. International direct-dialled telephone service is available to more than 200 countries and other overseas territories – covering 99 per cent of the world's 800 million telephones. In the year to 31 March 2000 BT's total turnover, including its share of its ventures' turnover, was £21,903 million, with a pre-tax profit of £2,942 million.

20

Corporate Reputation and the Internet

Simon Berkeley and Tom Woollard,
Environmental Resources
Management

Two new pressures face multinational corporations today: the global spread of e-commerce, and rising public expectations of corporate environmental and social responsibility. At first these seem unrelated, but our recent work points to some connections that large corporations will soon need to address.

New drivers of ethical performance for businesses

Consumer concerns

Concerns over the safety of beef destroyed a £7 billion industry in a few months, yet to date the link between BSE and CJD remains unproven. Shell lost substantial business in Europe over the Brent Spar incident despite environmentalists' endorsement of the decision. In MORI opinion polls on corporate responsibility over 90 per cent of those polled say environmental responsibility is important to them in choosing between products and services. In the past regulators have been the arbiters of ethical performance by business. Now increasingly, however, the general public and media are forming opinions and acting on them in defiance of government views (see Figure 20.1).

The Internet shifts the balance of power towards purchasers and consumers

Ford and General Motors plan to transfer all their purchasing to the Web in the next five years, allowing wider competition among their suppliers. Studies of the impact on business show a clear trend:

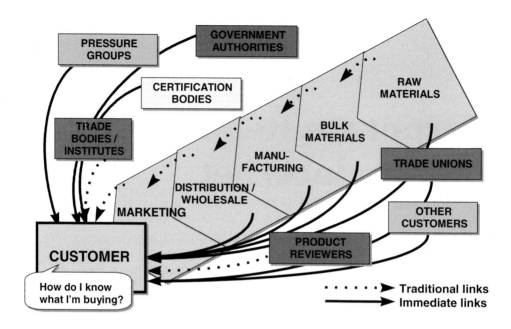

Figure 20.1 *Customers have access to a wide range of information and opinions on environmental and social impacts of products and services*

- 'The internet could be a strong price-deflating mechanism' (*Economist*, February 2000).
- 'The new economy has created such downward pressure on prices that it is safe to say inflation is dead' (*Harvard Business Review*, December 1999).
- 'E-commerce businesses are tilting their affiliation away from suppliers towards the consumer' (*Harvard Business Review*, December 1999).

The Internet is changing the landscape of competition; as price and location become less important firms will need to differentiate products and services on other attributes.

Campaign groups have the edge in the competition for credibility

Shareholders in BP are actively canvassed by Greenpeace in its campaign against the expansion of oil exploration. They are invited to explore the 'Sane BP' website, which provides detailed, factual accounts of the impact of exploration in sensitive areas. The Norwegian oil major Statoil is similarly attacked with a well-researched case for investment in renewable energy (www.newstatoil.com). An Internet campaign against Coca-Cola persuaded the company to change its mind about the use of HFCs in its refrigerators at the Sydney Olympics (*Financial Times*, 18 July 2000). Protest groups now mount

more effective campaigns against industries and companies on their environmental and social performance. Campaign websites are often more credible
than the bland assurances of corporate PR because campaigners focus on the
evidence of impacts while corporate websites focus on the *process* by which
impacts are managed.

Websites more credible than television?

A recent survey in the US found that people find websites more credible than
their traditional counterparts. Of those polled 54 per cent gave CNN.com a
high believability rating, while only 40 per cent gave the same rating to CNN.
What this may suggest is that companies will need to work even harder to
demonstrate the sincerity of their environmental and social practices to
business customers, investors and consumers. The Internet makes it far easier
for these key stakeholders to find out just how sincere they are (see Figure
20.2).

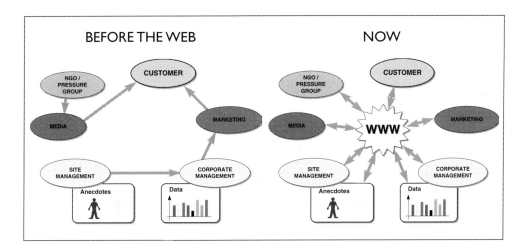

Figure 20.2 *Corporations now have less control over perceptions about them than in
the days when newspapers and television were the main media*

Issues ignited on the Web can 'catch fire'

Protests against the WTO meeting in Seattle in March 2000, and against 'global
capitalism' in London on 1 May 2000, took hold without the involvement of
any single co-ordinating body. In Seattle the meeting was successfully
disrupted, and in London the damage to property was estimated at £500,000.

Investors are turning to the Internet to assess companies' credentials

A recent survey of private investors in the US found that they are increasingly likely to go online for investment advice (*see* www.people-press.org). MediaWave, a successful five-year-old investor relations company, has recently launched a service to broadcast price-sensitive company information on the Web. A recent study of pharmaceutical companies finds that those with the best environmental records outperformed the S&P 500 in 1999 by 15 per cent (www.responsibilityinc.com). Investors researching on the Internet could well be influenced by views on companies' ethical performance.

Leaking corporate 'fire walls'

Leading companies we work with are finding that employees and contractors exchange confidential company information and views on Internet chat rooms. Often there is no malicious intent – the Internet is simply a convenient medium for communication. But this trend is leading some companies to wonder if information 'fire walls' are sustainable.

The response from business

We surveyed the websites of the FTSE100 companies to assess how well they convey their commitment to managing their environmental and social aspects (see www.erm.com). While many declare good intentions on environmental matters few use their Web-based communications tools to engage stakeholders (including customers) in a two-way dialogue on these issues. It is clear that few companies are exploiting the potential of the Internet to enhance their social and environmental reputation.

Our survey found that:

- Only 45 of the FTSE100 sites featured an environmental section.
- Very few websites (3/100) actively pursue two-way dialogue on environmental/social issues. Only four have links to lobby group sites.
- Very few FTSE100 companies reveal any information relating to their position on social/ethical issues (4/100).
- Protest websites are often more focused and clearer than the corporate sites they target.

Addressing the issues

Our studies, and discussions with corporate affairs and environment senior managers, suggest a number of approaches to deal with the new drivers for ethical performance (see Figure 20.3).

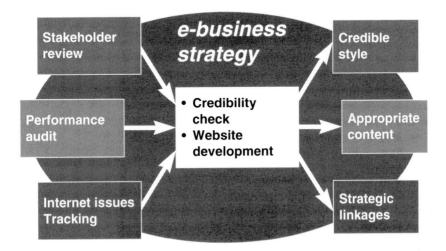

Figure 20.3 *Ethical positioning is a subset of your overall e-business strategy. There are several good places to start, developing on your current position*

Companies need to be clear about 'who they are', and communicate this in an accessible way

This sounds like a platitude, but many firms are finding that it is necessary to be clear about what business they are in before they think of transferring business processes to the Internet. Asking 'Who are we?' also provides a basis for identifying where we can be most effective in tackling environmental and social issues.

Companies must act to establish and raise credibility

BP invites visitors to their site to 'Ask John Browne' about anything that concerns them, and promises a reply a day from the Group Chief Executive himself. Shell's site invites specific feedback on a wide range of issues. Several major manufacturing companies publish on their website contact details for the environment, health and safety managers at their major sites (eg see www. daimlerchrysler.com/index_e). Companies will need to take a more open, honest, even humble approach to communication, and use performance data, targets and links to protest websites to build credibility.

Companies will need to address issues in terms that pressure groups use

For example, an emerging issue for the pharmaceutical industry is its lack of attention to provision of cheap drugs for the Third world. Yet this receives scant attention on most pharma-company websites. Biotechnology firms are, however, now starting to defend their industry with websites that mimic the protest sites (see www.cp.us.novartis.com/webackbiotech_frame), but the protest groups remain sceptical.

Corporate websites need to address multiple constituents using smart website design

Less well-informed visitors need information that tells a story they can understand. Opinion formers such as news media, protest groups and regulators need performance data and targets. These need to be addressed simultaneously, and leading companies now use smart website design to address this (see www.ubs.com and Figure 20.4).

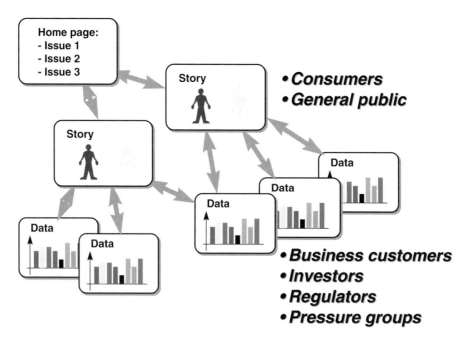

Figure 20.4 *Corporate websites need to address multiple constituents using smart website design*

Companies should consider forming alliances and Web links to reflect intentions and demonstrate action

Firms should get in touch with green groups, protest groups, etc, and collaborate on tackling issues.

Companies should use the Internet to establish better communications up and down their supply chains

Since the controversy over GM food, supermarkets now have better channels of communication with their suppliers on ingredients and processes.

Companies will need to scan the Web for emerging issues

To be of value, however, issues tracking needs an overlay of professional analysis to 'sort the wheat from the chaff'.

Dr Tom Woollard is head of ERM's Corporate Advisory Services (CAS) group based in London. Tom works with major FTSE100 companies on environmental reporting and management systems assignments.

Simon Berkeley is a Principal Consultant in the CAS team who specialises in strategic assignments including corporate reputation risk and performance indicators. He worked in the oil business in the aftermath of the Piper Alpha disaster, developing HSE Management Systems to help oil and petrochemical companies control risks in a complex business environment. After a period with Shell, he worked with Gemini Consulting on major strategy and change programmes for global multinationals before joining ERM's Corporate Advisory team.

ERM
8 Cavendish Square
London W1M 0ER

tel: 020 7465 7200
fax: 020 7465 7272
website: www.erm.com

ERM is a global environmental management consultancy with 2,500 staff working in over 34 countries. It is dedicated to supporting businesses, governments and international institutions in managing environmental and social issues. Its work for businesses covers strategic support in developing strategies for the environment, social issues or sustainable development as a whole. It provides a wide range of business and technical services to help firms implement these strategies, including performance management support, management systems development and certification, impact assessments, audits, special studies and issues tracking.

Section VI

Environmental Management Systems

21

Bite-sized Chunks of EMS for Smaller Firms

Matthias Gelber,
14000 & ONE Solutions

Challenges of EMS implementation for smaller companies

Small companies are facing increasing pressure from regulators and suppliers to improve their environmental performance. However, pressures associated with time, resources and competence often lead to many small companies putting the environment low on their agenda. Few small companies have investigated the inherent business benefits of operating an effective EMS, which could help them avoid inefficiency and enhance legal compliance.

Inertia within such small firms is extremely difficult to overcome due to various obstacles and barriers associated with implementation, including:

- perceived lack of time
- lack of policy and documents
- lack of understanding of EMS requirements
- lack of sufficient regulatory and legal knowledge
- economic constraints
- difficulty in defining economic benefits.

Project Acorn

Empowerment – Engagement – Partnerships
The Project Acorn Approach

'The only way to eat an elephant is in small manageable bites'

Project Acorn uses a phased EMS implementation approach that breaks down the internationally accepted standard, ISO 14001, into five levels. An optional sixth level allows companies to develop systems for public reporting and accountability, with the possibility of registering under the European EMAS Regulation. BSI and 14000 & ONE Solutions deliver the pilot project, with support from the Department of Trade and Industry (DTI) and DETR.

At each level of the EMS implementation individuals receive training so that they can develop their company's system and environmental performance indicators. These will allow them to track and report (internally or externally) on their company's environmental performance over time, and benchmark performance in critical areas.

After each level of the scheme has been implemented the company is assessed to ensure that the requirements of each level have been met. If the company is successful it is awarded a certificate of achievement so that it can demonstrate progress to its key customers and other interested stakeholders. At the end of level 5 an audit will be conducted to ensure that the company's EMS meets all of the requirements of ISO 14001. If successful the company receives an ISO 14001 certificate.

If a company progresses to level 6 then the external audit will verify the data and the system against the requirements of EMAS.

What are the six levels of the scheme?

Project Acorn provides a staged route to ISO 14001 certification and/or EMAS registration. In addition, it is one of the aims of the pilot project to develop environmental performance indicators to reflect the environmental aspects of activities, products or services. Project Acorn will focus on resource efficiency. Smaller companies will be motivated to drive forward the environmental agenda if there is the opportunity to contribute to the bottom line and thus enhance senior management support.

The advantage of the staged approach is the ability for smaller companies to get recognition for the intermediate achievements of progress while implementing ISO 14001. Making an EMS work within those companies needs significant time in order to tailor the system to their needs and for them to gain real ownership of it. Smaller companies appreciate clear guidance on every step they need to take on the journey of EMS implementation and can learn from the experience of other companies. The staged approach will avoid the waste of misdirected resources as there will be periodic checks of progress.

The six levels of the model are:

1. Top management commitment to the project, conducting a baseline assessment.
2. Legal, customer and market requirements.
3. Confirmation and management of significant environmental aspects.
4. Launching an effective EMS.
5. Checking, audit and management review.
6. Data verification, public reporting and EMAS registration.

These levels are very closely aligned with the logical progression of an ISO 14001 implementation project. The inclusion of continual improvement and culture change from day one will immediately ensure appropriate levels of ownership and performance focus. This will give smaller companies systems that work and which deliver performance improvements.

Working for shared goals

One of the main drivers of Project Acorn is the supply chain relationship element. Many larger companies that support Project Acorn have been involved in environmental initiatives for some time, often resulting in third-party certification to ISO 14001 and/or registration to EMAS. However, as their system matures it becomes apparent that the environmental impacts of the products they manufacture or the service they deliver are often rooted in the components, materials and services supplied to them by other companies within their supply chains. These larger companies therefore have a vested interest in improving the environmental performance of their supply chains. Some of the mentors participating in Project Acorn are Severn Trent, Vauxhall, Wessex Water, CGNU, Cable & Wireless, BT and ASDA.

Project Acorn facilitates communication through a customer–supplier partnership addressing the environmental concerns of all parties. The Project Acorn approach has received recognition from participants because it offers practical support for smaller companies, facilitates supply chain communications and leads to improvements for the mentor company. Environmental management that contributes positively to the business relationship between suppliers and large customers will carry significant weight and motivation. For mentors, Project Acorn will support their goal of supply chain management and the contribution to corporate sustainability goals.

EPE and performance indicators

Development of the new international standard for environmental performance evaluation (EPE), ISO 14031, has added impetus to an already growing realisation that systematic and creditable performance measurement is a necessary element if environmental management is truly committed to continual improvement.

The application of ISO guidance on EPE will help add value to Project Acorn businesses by enabling them to track their resource consumption, compliance with customer requirements and by setting SMART targets to key performance indicators.

Wider application

Project Acorn aims to turn an obligation to environmental management into an opportunity for your business by:

- empowering your organisation by giving you the training, tools and ownership;
- engaging you by providing the right conditions for smaller companies;
- partnership with you, your customers and Project Acorn.

The Project Acorn vision is a model that improves environmental performance with added value for those who participate. In the future more mentors and suppliers will be using the Project Acorn model, and other delivery agents and certification bodies will make the approach widely available in the marketplace. Telephone helpline and Web support will be available. If you are interested in becoming part of the current Project Acorn pilot, please contact: Fiona Gibbons, Project Co-ordinator, contact point at BSI for Project Acorn. Tel: 020 8996 7665; E-mail: Fiona.Gibbons@bsi.org.uk

Matthias Gelber, Technical Director of 14000 & ONE Solutions Ltd, is an internationally recognised expert on environmental management systems and environmental performance evaluation. He acts as head of the International Network for Environmental Management Delegation at ISO TC207 (SC1, SC4). He is also a member of the UK British Standards Institution standards committee on EPE.

Matthias, was involved in an EU project on preventive environmental management for small and medium-sized enterprises and is currently working on Project Acorn as technical specialist for EMS and EPE implementation and training.

14000 & ONE Solutions Ltd
PO Box 1005
Stoke-on-Trent
Staffordshire ST1 3TN

tel: 070000 14000

fax: 070000 14001
e-mail: mgelber@14001.com

14000 & ONE Solutions Ltd is a leading edge environmental consultancy with worldwide experience on ISO 14001 implementation and training. The company also provides advice on waste minimisation, corporate environmental strategy, environmental reporting, supplier assessment and evaluation, environmental purchasing, ISO 9001 (2000) and OHSAS 18001.

14000 & ONE Solutions Ltd has been responsible for the first UK pilot application of ISO 14031, which has been funded by the DTI. 14000 & ONE Solutions is now responsible for the development of technical methodology and delivery of Project Acorn, the UK's largest staged EMS implementation programme.

The company has developed an excellent reputation for pragmatic advice and training, having delivered courses and consultancy in over 20 countries and to a wide cross section of industrial sectors.

22

From EMS to Responsible Care Management Systems: A Logical Transition

Frank Richardson,
Thomas Swan

Historically, banks and insurance companies have always complained about having to pay for industry's mistakes. Gradually, these and other stakeholders are beginning to realise that management systems have an important role to play in industry – as risk reduction tools.

And it's not just about obeying the law. To be fair, industry has been obeying the law for decades. Nevertheless its public image is generally poor. Doing better is all about going the extra mile. The benefits can be significant, especially for your company, and important for the environment as a whole.

So, what is the 'extra mile'? Well, it's certainly not one factor alone but a combination of factors such as: an in-depth analysis of how your activities affect the environment (including employees), improvement programmes, regular audits (not just of your own activities but also those of transport contractors, on-site contractors, suppliers, customers), sharing your problems with other companies, learning from their experiences and practices, getting your own employees more involved (there's valuable experience out there to be tapped). Interestingly, these additional factors come under the umbrella of 'responsible care' (the Chemical Industries Association (CIA) initiative towards sustainability).

By the end of 1996 Thomas Swan was registered to ISO 9001, ISO 14001 and EMAS. As a member of CIA the company was signed up to responsible care, which includes both occupational health and safety (OH&S) and environmental (OHS&E) requirements. But responsible care also asks for more than what the relevant standards (ISO 14001 and OHSAS 18001) require, the key additional requirement being product stewardship. Product stewardship goes beyond what happens to your products/services on site – you must also be aware of, and try to influence, the performance of your customers, suppliers and hauliers, etc, in terms of both the environment and occupational health and safety.

Having an environmental management system already well established gave us an excellent foundation on which to either add another system addressing OH&S or take the opportunity to develop a fully integrated management system to address OHS&E and responsible care.

So, when we started the development of a management system to BS 7750 (later ISO 14001) in 1992, we already had tentative plans to extend the system to comply with EMAS, the expected introduction of an OH&S standard (BS 8800, now OHSAS 18001) and then, finally, to integrate everything under a responsible care management system (RCMS), to be externally verified by a certification body. This was achieved in December 1999.

What are the essential ingredients for achieving a certifiable responsible care management system? Probably the most important is genuine commitment from the most senior level of management, but teamwork and patience are also very important.

Teamwork, and having the right team, are essential. Ideally you need broad, company-wide representation. This will give you the experience and expertise you require as the exercise progresses. Representation from engineering, production, technical, sales, purchasing and quality should be your minimum target. Other advantages of broad representation are that it helps the company as a whole to feel involved, and ensures that the system becomes a natural part of day-to-day management.

Thomas Swan was registered to BS 7750 in 1995, and both ISO 14001 and EMAS in 1996. At this stage of the exercise we had to decide (as will many other companies) whether to develop another 'stand-alone' management system complying with the new BS 8800/OHSAS 18001 to run alongside our ISO 14001 system or develop an integrated system. We decided on the latter course. When you read the two standards it is obvious that the various elements are readily open to integration. Indeed, for some elements it is more difficult to segregate environment and health and safety than to integrate them, for example:

- accident investigation;
- risk assessment;

- auditing;
- improvement programmes.

To add responsible care requirements to the integrated system it was necessary to conduct a 'gap analysis' to identify those extra requirements of responsible care not covered by the standards. Our own gap analysis identified the following as some of the major new or more in-depth requirements:

- emergency preparedness;
- training for personnel outside of the organisation;
- sharing of relevant OHS&E knowledge and learning from the experience of others;
- purchasing;
- contractors;
- equipment (including maintenance and inspection);
- packaging and labelling;
- hauliers, storage contractors and distributors;
- management of change.

As stated earlier, the task of developing an integrated RCMS starting from an EMS isn't difficult, but it is time-consuming. Remember, the amount of legislation already affecting any company can provide a good start to a management system. Examples of such legislation are:

- IPC (IPPC);
- HASAWA;
- CIMAH (COMAH) Regulations;
- Management of Health & Safety at Work Regulations;
- Noise at Work Regulations;
- Electricity at Work Regulations.

Once you have your system certified, however, is that the end of the story? Absolutely not! A key feature of the standards is continual improvement. Basically this is achieved by first taking a hard look at how your business affects the OH&S of your workforce and the wider environment.

- Where there are problems – do something about it – and keep doing it!
- Where there are benefits/advantages – don't stand still – do even better!

Some companies have an inbuilt fear of 'management systems'. As we have already stated integration need not be difficult, but:

- Will the exercise cost money?   Yes
- Will it involve lots of people?   Yes
- Will it take time?   Yes
- Will we encounter many problems?   Yes
- Will we now be subjected to inspection visits?   Yes
- Will we have to measure our performance?   Yes
- Will we have to conduct our own audits?   Yes
- Will the public know more about our activities?   Yes
- Will it all be worthwhile?   Yes! Yes! Yes!

Finally, the essence of a responsible care management system is risk assessment, risk assessment, risk assessment!

Frank Richardson is the Technical and Quality Manager at Thomas Swan. He spends some 80 per cent of his time on research and 20 per cent on quality, and yet with the full support of the managing director and board has led the company through the development, implementation, registration and maintenance of the management system standards (quality, environment and occupational health and safety).

Thomas Swan & Co. Ltd
Crookhall
Consett
Co. Durham DH8 7ND

tel: 01207 505131
fax: 01207 590467
e-mail: enquiries@thomas-swan.co.uk
website: http://www.thomas-swan.co.uk

Thomas Swan & Co Ltd is a family-owned chemical company with around 160 employees based in north-east England. The company produces a broad range of fine and performance chemicals with applications in the polymer, paint, rubber, ink, adhesive and pharmaceutical industries, exporting in excess of 70 per cent of its products worldwide. The company is also involved in custom synthesis and contract manufacture.

Thomas Swan has developed a reputation as a reliable and successful business partner and is a leader in the adoption and implementation of quality, environmental and occupational health and safety standards, including responsible care.

23

The Future of Environmental Management Systems: Web-based Applications

Agneta Gerstenfeld and
Hewitt Roberts,
Entropy International

IT Solutions for Environmental Management

Environmental IT is a fast-moving sector. In the last few years we have seen an explosion in the software market, and the tools available range from relatively simple single-user data tracking software applications to highly complicated custom-made and networked environmental information management systems. However, stand-alone, single-user software applications are yesterday's news. Businesses today are looking for network applications to more effectively exchange information with customers, partners and employees worldwide and thus trying to implement solutions that take advantage of existing investments in skills and technology.

The late 1990s and even the early days of this new millennium have been synonymous with rapid advancements and worldwide uptake in the use of the Web, Web technologies and Web applications. While much is heralded about the usefulness of the Web for personal use, headlines in the business press and stories of outstanding corporate success are regularly associated with the application of Internet, intranet and Web technologies.

With Web applications numerous computers can be connected in a network to a central server-based software application accessed through a Web browser. Web applications can simplify business processes such as reporting, project management, purchasing, inventory management, marketing, sales, and of course environmental management.

So what is all the hype about? In short, Web technology is changing the way companies do business. Organisations that adopt this new technology, whether it be for corporate-wide communication or even online environmental management, are enjoying clear business advantages by becoming more efficient, more responsive and ultimately more competitive.

Benefits of Web technology for improved environmental management

Web applications on the Internet or on a corporate intranet provide for real-time data input and reporting, they allow multiple users simultaneously to access and use a common software tool. They promote collaboration, improve the way people carry out important business processes and ensure that all users are getting accurate, timely information at the lowest possible cost. Clearly this enhances business efficiency, improves management practices at all levels of an organisation and ultimately enhances corporate survival.

The real power of Web technology is in the use of applications for improving day-to-day management practices and, similarly, intranets that provide for improved corporate environmental management.

Whether it be a simple tool for company-wide environmental reporting or data collection to a fully functional ISO 14001-compliant online environmental management system, intranets are similarly becoming the solution of choice for improved environmental performance and will similarly change the way companies manage their environmental affairs and catalyse efforts for improved environmental management.

Four of the most obvious benefits of an intranet or Internet-based environmental management solution are:

- time savings;
- cost savings;
- improved knowledge sharing and retention;
- improved management practices and consistency.

These benefits are clearly seen in a case study comparing two companies (Company A and Company B) that environmental software specialists at Entropy International have recently worked with. Both companies needed help to develop certifiable environmental management systems. Both companies are multi-site and multinational. However, Company A chose to develop a paper-based environmental management system at each of its sites and Company B chose to implement Envoy EMS, an online, intranet-based environmental management system.

Envoy EMS is an award-winning EMS solution developed by Entropy International and has been designed to streamline and improve the management of environmental, safety and health issues for single-site and multi-site corporations. The system is designed to provide a step-by-step template for implementing and documenting an ISO 14001 certifiable EMS and a dynamic tool to plan improvement, track action and implement change.

In the time that Company A built management systems at four sites, Company B will have built EMS at 185 sites (see Figures 23.1 and 23.2).

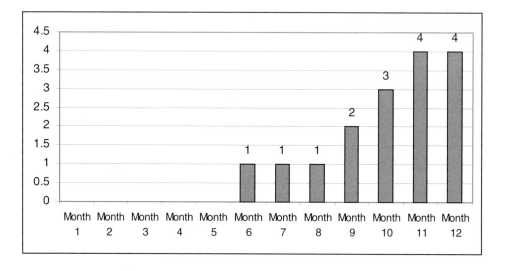

Figure 23.1 *Number of EMSs developed – Company A*

In addition Company B worked more than 46 times faster than Company A, due to the economy of scale realised from the use of an intranet. It also drastically reduced the per-site price of their EMSs. As can be seen in Figure 23.3, the cost of EMS development for Company A was over £10,000 per site compared to that of Company B's £560 per site.

The advantages for Company B don't stop there. The aspect of knowledge retention and a growing 'knowledge base' are of considerable value in the case of Company B. Clearly, the knowledge and experience Company A gained from EMS development rests with the individuals that developed the EMSs on each of their sites. Consequently once they are gone so is that experience

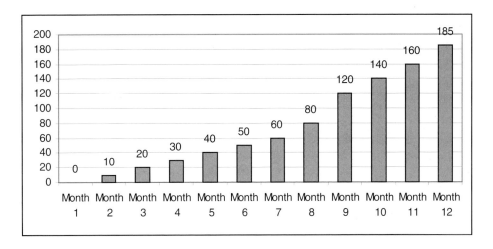

Figure 23.2 *Number of EMSs developed – Company B*

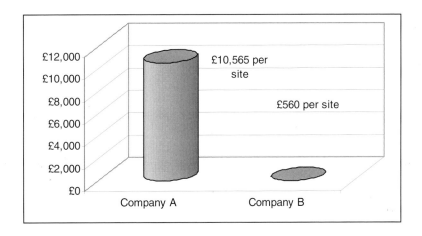

Figure 23.3 *Average cost per EMS*

and knowledge. However, and as is the nature of database-driven intranet applications, the knowledge both of EMS development and of all the communi-cation, queries, answers and best practices recorded over time are retained within Company B's intranet-based EMS application for easy access and indefinite use.

Although the benefits of using an intranet application are wide and varied a fourth clear benefit is that of obvious site-level management system con-sistency and improved overall management efficiency. Using an intranet-based EMS template common to all sites means that all sites understand each other, can help each other, can report and compare results easily and ultimately speed and streamline the process of management system development and administration.

Case study

Implementing Envoy at Aggregate Industries, UK

When Aggregate Industries UK Ltd launched its corporate-wide ISO 14001 initiative, aiming to certify all its sites in the UK before the end of 2001, the Environmental Working Group soon realised the benefits of utilising Envoy. By doing so they could save time, save money and leverage and simplify the environmental management effort within the company. For Aggregate Industries Envoy was an action-oriented solution to help it rapidly improve environmental management practices throughout the company. Envoy provided a cost-effective, user-friendly and extraordinarily powerful management solution as well as a common corporate framework for ISO 14001 implementation.

> Having accepted the fact that we needed a software solution for improved environmental management, we investigated many packages. Most available tools seemed to miss the mark on one or more issues. Envoy EMS was the best solution for us and it took little time to convince our directors that it was the only way forward. Coupled with unparalleled support from Entropy International, we are sure that we established the best possible future for our EMS.
>
> (Miles Watkins, a driving force behind the Envoy project at Aggregate Industries)

A tool to identify significant environmental aspects

Envoy EMS provided a tool for identifying, assessing and documenting the significant environmental aspects and impacts of all sites in the organisation. The Environmental Working Group worked with Entropy International to refine the module to reflect the company's own requirements and through the self-establishment of the significance level it has formed a timesaving, effective device. Once the significance assessment was complete, a comprehensive and completely up-to-date Register of Environmental Aspects was automatically generated.

(Miles Watkins)

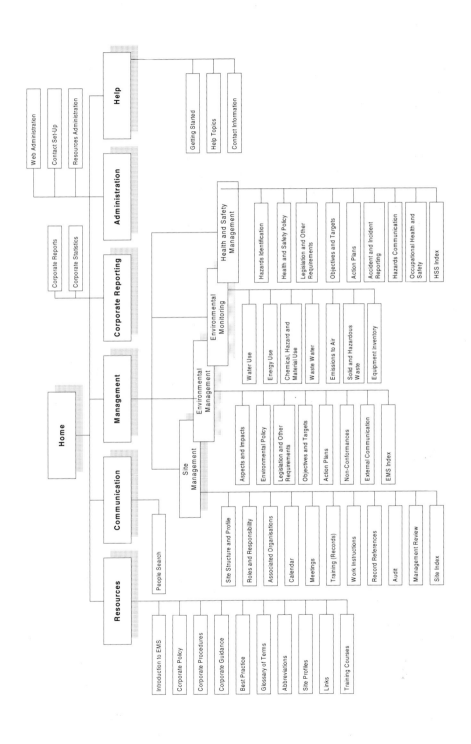

Figure 23.4 *Envoy EMS: the system*

A tool for improved environmental performance

Setting, follow-up and evaluation of objectives and targets, action plans and non-conformances are central features of an environmental management system. In an intranet-based system this is easily done. When asked to evaluate the performance of the system the Environmental Working Group commented: 'Owing to our single certificate registration to ISO 14001, we need to track the site management programmes centrally. Review and target dates are quickly accessible in Envoy EMS and therefore we can spend more time on environmental improvements and less time on the phone.'

To Miles Watkins, the single store of data is the key: 'With upwards of 140 sites and only six environmental staff the effective management of information is critical. Envoy EMS provides a consistent framework into which the essential data can be entered and accessed by all with zero time delay.'

Documentation and document control

Envoy EMS is compliant with ISO 14001 and the standard system structure worked well for Aggregate Industries. According to Miles:

> a few customisations to the interface were made and our existing EMS documentation was simply cut and pasted into the system. This solved the problem of transition from a paper-based system to Envoy and reduced the time spent on administration. And given that document control was our biggest nightmare, with Envoy EMS there is only one set of controlled documents and they are stored centrally; a huge saving in time and money.

Envoy EMS as a reporting and communication tool

One of the main advantages of an intranet-based application such as Envoy EMS is the instant access to information and the provision of reporting and communication features. Standard and customisable reports make it easy to evaluate any part of the environmental management system. Additionally, corporate-level users can generate reports for one site or all sites in the system as well as instantly generate 'real-time' corporate-level statistics showing environmental performance across all sites. According to Miles the system:

> provides a method of finding answers quickly. The numerous reports that the package includes means that we can rapidly assess where we are on any number of issues. The biggest time savings are to be had with the tracking of non-conformances and the communication module that allows easy access to people with specific duties related to them.

A complete EMS solution

Envoy EMS is the first system of its kind, and given the increasing presence and applicability of corporate intranets it is on the cutting edge of technology and environmental management. To proactive companies such as Aggregate Industries it is obvious that this is the future of management software applications.

Agneta Gerstenfeld is Project Co-ordinator at Entropy International responsible for EMS implementation projects and training.

Hewitt Roberts is a director of Entropy International. Hewitt has authored and co-ordinated numerous environmental management projects at national, European and international levels, including EMS implementation, multimedia training, environmental management information systems development and design.

Entropy International Environmental Consulting Ltd
96 Church Street
Lancaster LA1 1TD

tel: 01524 389 385
fax: 01524 389 386
E-mail: info@entropy-international.com
website: http://www.entropy-international.com

Entropy International provides information technology-based business management solutions that improve business efficiency and promote and improve corporate environmental performance, working environment and quality.

In 1999 Entropy International won the British Telecom Businessconnections Award for Innovation, as a result of their work with Joseph E. Seagram and Sons and the Envoy intranet-based environmental management solution.

24

Certification and Verification: A Legitimacy Issue

Christopher Sheldon,
BSI Global Quality Services

A good environmental management system should be able to go through more cycles than the Tour de France. Certification of an EMS by the British Standards Institution (BSI) helps to improve the breed by ensuring all the benefits get through to the bottom line.

It's almost a cliché to say 'certification is not a destination – it's a continual process', but such a saying proves why clichés are so difficult to put down once they've been taken up. They stick like Velcro ™ to our mental concepts, and what holds them there is the strong element of truth woven into their strands. Too often client managers from certification bodies are confronted with company managers who wax lyrical about commitment before they have gained a certificate to ISO 14001, but whose interest in retaining it wanes with each passing year.

Unsurprisingly, surveys show that most busy managers are driven by fear of legislation or the marketplace to implement an environmental management system. There's nothing wrong with having a healthy respect for either of these forces for change, but when they move from being contributing factors to become a company's sole motivation the only one to lose out is the business itself. Avoiding this '14001 as licence to trade approach' is where certification can add value to any company, at any stage of its EMS development. These benefits can even start during the process of implementing prior to formal recognition.

One of the advantages of having an independent and objective specialist in EMS look over your growing system lies in the expertise he or she can bring in the form of industry best practice. While stopping well short of consultancy this can be particularly helpful during the development of a new system. Phased assessment processes mean that companies can concentrate on fundamental areas of concern (impact assessment and identification of relevant legal requirements) before moving on to build their EMS on secure foundations.

It's also not unknown for some companies to have run an EMS for many months before seeking certification, convinced that they are already meeting the requirements of the standard. These same companies frequently find that following a certification visit they have a surprising number of non-conformances. The major 'hot spots' for new systems are outlined in the box at the end of this chapter.

With all these advantages built in prior to gaining an ISO 14001 certificate, it's no surprise that many executives forget to look beyond that achievement, and consider what they could gain from the continuing assessment visits. Most regard such visits simply as a way for the certification bodies to check up on whether the company's EMS still meets the requirements of the agreed standards. This is a reductionist view bred by 'command and control' cultures, and limited application of ISO 9000 systems; seeing beyond it can open up new vistas.

The hard facts of the market are that if certification bodies do not add value to a client's system the client will simply go and find someone who can. Yet talk of a value-added approach to environmental management can lead to the assumption that we all know precisely what value we are talking about. Is it measured in money, well-being, or air quality? The answer, in the case of ISO 14001 registration, is all three. The benefits that lie in the continued maintenance of an EMS can be equally significant, and equally measurable:

- **Certification process distils industry best practice** – Continual improvement is a principle built into much of the UK's environmental legislation. Exceeding legally imposed performance requirements takes knowledge of the best tools and techniques available. Other than regulators, no one has a better overview of best practice than a certification body.
- **Solid foundation for verified environmental reporting** – Whether compiling a report to meet the requirements of the revised Eco Management and Audit Scheme (EMAS II) or the Global Environmental Reporting Initiative, facts, figures and raw data will only gain credibility if produced by an EMS that is well maintained and tightly focused. Certification underpins that credibility.
- **Stakeholder security** – A company may have many stakeholders, but one badge is recognised by all of them. Evidence suggests that certification allows a growing confidence in the dialogue between stakeholder and

organisation to remain undented, even when threatened by adverse environmental factors.

• **Acknowledged by regulators** – Even the Environment Agency itself considers that the benefits to be gained by ISO 14001 do not stop short of certification. Though public accountability and the government watchdog, the National Audit Office, have certainly added their weight to the 'rolling out' of 14001 in the Civil Service, the Agency has had more chances to scrutinise the difference between successful and unsuccessful management systems. Its decision to go for certification acknowledges that the process has a definite place in terms of meeting public accountability for environmental management.

• **Demonstrable improvement in environmental performance** – Organisational progress against targets is an integral part of an EMS. Continued certification against ISO 14001 provides objective evidence that an organisation is delivering against its own goals. The organisation itself decides how much detail it wishes to add beyond the retention of certification.

Interestingly, these advantages are not simply hearsay but measurable outputs of healthy management systems. Using ISO 14031's interconnected system of management and operational performance indicators, coupled with environmental condition indicators, the certification process can be directly linked to bottom-line benefits. The process and indicators will vary with an organisation's individual circumstances and needs, but the appropriateness of the chosen indicators will be covered by the assessment against ISO 14001.

At best, certification can help companies to avoid the 'dumbing down' of their management systems. The novelty value of new systems is quickly outdistanced by the benefits they bring but, as inevitable as the ageing process itself, familiarity can breed contempt as a series of managerial blind spots. Environmental objectives that started out as stretch goals become elements on a familiar landscape, to the point where potential bottom-line gains are simply overlooked. 'We've done the environment, haven't we?' becomes the accepted opinion.

Successful companies know that the periodic overview by an impartial yet experienced third party restores the original impetus for gain, brings fresh energy to harness further profits from change, and tests them against the best. Not bad for a framed piece of paper.

HOW CERTIFICATION REVEALS EMS HOTSPOTS

Common problems with EMS, flushed out by ISO 14001 certification

- **Initial review** – Often the toughest part for a company because it usually lack expertise in-house. Certifiers can show how the review has become a self-contained exercise that companies find hard to relate to their everyday activities.
- **Significant impacts** – Frequently no prioritisation or weighting has been ascribed to the individual environmental impacts, and the impacts related to service activities are typically overlooked. Certifiers can spot gaps and can assess effective methodologies for rating one impact against another.
- **Legal and other requirements** – Many companies bring in legal professionals to help identify what legislation applies to them, or use a commercially available legal database. Certifiers know this can still leave the problems of identifying and disseminating appropriate information unresolved.
- **Objectives and targets** – In many cases the objectives and targets not only reflect the cash-flow priorities of the company, but create the problem of allowing cost-benefit analysis to prioritise the impacts, not the objectives. This can lead in turn to fundamental problems with the EMS, and certifers are better diagnosticians than stakeholders in this area.
- **Environmental management programmes** – The most common problem here is the identification and subsequent ownership of new responsibilities. Certification shows that older, more established work patterns and practices can often displace them.
- **Documents and their control** – This is the biggest bugbear of all for companies, especially as many equate objective evidence of management processes with a document almost by default. Documents are not the only proof, as certifiers will happily discuss.
- **Operational controls** – Not all controls are successfully cross-referenced with the requirements of the objectives and targets; the result can be operational gaps that will allow problems to develop unseen until too late. Certification can help to spot such gaps.
- **EMS audits** – An EMS can only be as effective as its audits. If a company is suffering from ineffective internal auditing it is usually the last to know. Only external testing of the system concentrating on root cause analysis and relating evidence to findings can reveal this to company management.
- **Management review** – The difficulty of sustaining support and commitment for an EMS once the system has achieved the early savings can dog many reviews. Certification can check that strategic input remains timely and appropriate.

Christopher Sheldon is currently working with BSI in Global Quality Services on a number of environmental initiatives. He is an international policy adviser, trainer, author and broadcaster on environmental management. He has been involved in the highest levels of EMS development for over a decade through standards institutions, professional bodies and commercial application. He now acts as a consultant in the areas of management systems and related sustainable development issues.

The **British Standards Institutution** (BSI) is the world's largest national standards body and in 2001 celebrates its centenary. As a founding member of the International Standards Organisation (ISO), BSI has over 35,500 registrations to management standards worldwide. It offers registration to, and training and consultancy in, standards such as ISO 14001 – the International Standard for Environmental Management.

British Standards Institution
389 Chiswick High Road
London W4 4AL

tel: 020 8996 9001
fax: 020 8996 7001
e-mail: info@bsi-global.com
website: www.bsi-global.com

The ISO 9000:2000 series comprises four specific standards. ISO 9001:2000 itself is the core standard:

- ISO 9000:2000 – Quality Management Systems fundamentals and vocabulary
- ISO 9001:2000 – Quality Management Systems requirements
- ISO 9004:2000 – Quality Management Systems guidelines for performance improvement

Section VII
Energy Management

25

The Goods, the Bads, the Carrot and the Stick: Climate Change Levy

Judith Hackitt,
Chemical Industries Association

There are times when a philosophy of 'wait and see' is perfectly acceptable and sensible, but there are also times when it is a foolish strategy. Take climate change. Is it really happening? If so, how serious is it? Is this just a natural cycle? What are greenhouse gases anyway? Could this be just another (one of many) media scare story?

No one really knows the answers, but most of us are agreed that we simply cannot wait and see before we take action – the stakes are just too high. Changing our behaviour today to safeguard the future needs everyone's participation – encouraged and persuaded by the use of carrot and stick.

The government committed the UK to achieve a 12.5 per cent reduction in greenhouse gas emissions by 2010 in Kyoto, with an aspirational target of 20 per cent for CO_2. Industry should support the government's objective by committing to play its part for the environment and also because it makes business sense.

The UK chemical industry has some valuable experience to share and also views on what will, and will not, be successful in changing people's behaviour.

We are an energy intensive sector, with a total energy bill approaching £1 billion a year. We are also part of a global industry – 71 per cent of production is exported, 35 per cent outside of Europe. We are also important to the UK economy – employing 250,000 people direct, many more indirectly and contributing £4 billion annually to the UK trade balance. Staying competitive

Table 25.1 *Maintaining competitiveness is already difficult given energy price differences*

Chemical industry electricity prices as delivered pence/kWh (10–30 MW)					
Country	Reason for inclusion	1999 delivered price p/kWh	New energy/ green taxes p/kWh	Total p/kWh	Price relative UK=100
UK		3.06	0.30	3.36	100
Germany	Largest chemical producer in Europe	2.54	0.05	2.59	77
France	Second largest producer in Europe	2.60		2.60	77
Netherlands	Competing location	2.41		2.41	72
Belgium	Competing location	2.60		2.60	77
US	Largest chemical producer/exporter; world price setter for many products	1.95		1.95	58

is key to the success of our industry and energy prices are already higher than in competitor countries, especially the US (see Table 25.1).

Energy efficiency is part of our culture, as our track record shows, and we are committed to ongoing improvement. Between 1967 and 1990 we improved our energy efficiency by 60 per cent and since 1990 we have achieved further improvement. We have a voluntary energy efficiency agreement with government, committing us to a 20 per cent improvement from 1990 to 2005. We don't need an energy tax to change our behaviour – or to worsen our competitive position (see Figure 25.1).

When the Chancellor announced the climate change levy in the Budget in March 1999 we immediately voiced our concern and disappointment. The full levy would raise £145 million per year from CIA's member companies against a benefit in reduced employers' national insurance contributions of only £10 million.

The government has offered rebates to energy-intensive sectors that enter into energy efficiency agreements and we are, of course, negotiating an agreement to mitigate the impact of the levy on our members, but we believe there is a better way forward.

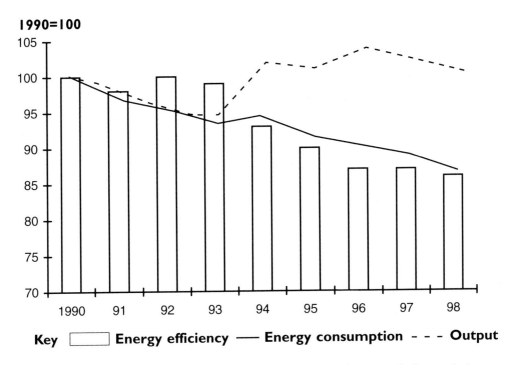

1990=100

Key ☐ **Energy efficiency** —— **Energy consumption** - - - **Output**

Figure 25.1 *Energy efficiency – CIA membership (based on matched samples)*

Aside from the fact that the levy applies only to business and not to the domestic sector, we believe the real opportunity, in a better-thought-out scheme, is to incentivise the whole of UK industry to play its part in meeting the UK's commitment to Kyoto. The current intention is to make the levy revenue neutral for business overall, but that neutrality should really be at the level of the individual firm, or at least sector, rather than transferring money from manufacturing industry to the service and public sectors. To make this possible the levy should not be linked to raising a fixed amount to fund a reduction in employers' national insurance contributions.

The purpose of the climate change levy should be to stimulate all companies to join demanding energy efficiency schemes, irrespective of how energy intensive the companies are. Much of the scope for improved efficiency lies in energy non-intensive sectors. Encouraging companies in all sectors to join schemes would promote a positive attitude to energy efficiency and encourage sharing of best practice, benchmarking with peers – an energy efficient culture. Any company, group or sector that is prepared to enter into an agreement should have the opportunity to do so provided that certain criteria can be met, including setting challenging energy efficiency targets and establishing auditing and reporting systems.

Companies entering into such schemes should be assured fiscal neutrality through exemption, rebates or some form of capping mechanism, but the climate change levy would apply in full to all business users that were not party to an agreement. A much more logical and fairer application of carrot and stick!

The current design of the climate change levy is justified by its proponents as being consistent with the government's commitment to shift the burden of taxation from 'goods' to 'bads'. But this assumes that energy use is 'bad' – it isn't. Inefficient use of energy – 'waste' – it bad; efficient, well-managed use of energy to provide the products and services we require is good!

Negotiated energy efficiency agreements are a good carrot; energy taxation should be the stick for the 'wait and sees'.

Judith Hackitt is Director of Business and Environment at the CIA. She graduated from Imperial College, London in 1975 with a degree in chemical engineering and joined Exxon Chemicals at Fawley as Process Engineer. She has held a variety of management positions including Technical Manager of Paramins and Operations Manager, Butyl Polymers. Judith joined Harcros Chemicals (now Elementis) in 1990 as Operations Director in the Pigments division and then become Group Risk Manager with worldwide responsibility for HSE and all risk matters in 1996.

The mission of the **Chemical Industries Association** (CIA) is to help members secure sustainable profitability and improve recognition of their contribution to society. It does this by working with them to influence relevant people and policies and by stimulating and helping them towards appropriate internal action, singly or co-operatively.

Chemical Industires Association
Kings Buildings
Smith Square
London SW1P 3JJ

tel: 020 7834 3399
fax: 020 7834 4469
e-mail: enquiries@cia.org.uk
website: www.cia.org.uk

26

How the Workforce Plays a Vital Role in Saving Energy and Reducing Waste: Perkins Engines

Uly Ma,
Energy Efficiency Best Practice
Programme

Examples of companies adopting energy saving and waste minimisation schemes are all around us. However, what makes the story at Peterborough-based Perkins Engines interesting is the fact that the initiative began with the workforce and grew to include active union participation.

Bottom-up approach

Perkins Engines manufactures a wide range of diesel engines at its Northamptonshire site. The company's management and workforce had been interested in reducing environmental impact for some time, but the visit by two of the firm's shop stewards to a World Class Manufacturing course in early 1996 proved to be the catalyst for a major savings initiative. The seminar, organised by the Amalgamated Engineering and Electrical Union and supported by the Energy Efficiency Best Practice Programme (EEBPP), was attended by union members Tony Ellingford and Paul O'Hearn. The session on energy saving and waste minimisation caught the shop stewards' imagination and they realised that improvements could be implemented at Perkins Engines.

Starting out

Although some management-led environmental initiatives already existed, none had generated sufficient enthusiasm among the workforce. Ellingford and O'Hearn set about devising a company-wide scheme for making improvements in energy efficiency and minimising waste. Knowing that total quality projects are encouraged by the management at Perkins, the two shop stewards decided to present their case to managers in this format. Colleagues were recruited into the scheme to lend support and the Director of Environmental Engineering was then asked to sponsor the initiative, which became known as Super Savers, because of the massive potential for saving energy and resources.

Communication is key

The success of any energy efficiency project depends largely on the way ideas are communicated to the stakeholders. At Perkins, Super Savers set out to communicate to the entire workforce that each employee had a part to play in improving environmental performance, and that good housekeeping can be carried out most effectively by the workforce itself. At the start of the campaign a staff survey showed that although there was a willingness to reduce energy consumption – 90 per cent said they were concerned, and that energy was wasted in their department – only 35 per cent took regular practical action. Fewer than 10 per cent of the workforce had been involved in company-driven environmental initiatives, citing lack of knowledge as the main reason. This pointed clearly to a need for better communication.

   Posters, information sheets and a newsletter, along with a handbook that was supplied to every member of staff, promoted Super Savers activities. The initiative was also included in existing quality and environment programmes running at the factory, in order to ensure that focus was maintained.

Rapid results

The initial objective, to raise workforce awareness of energy saving and waste reducing measures, was achieved very quickly. A year after implementation awareness of the company's environmental policy among the workforce had increased from 50 per cent to over 80 per cent. The number of workers who said they had done something practical to reduce environmental impact increased fourfold, to over 80 per cent. Perkins estimates that Super Savers has actually saved up to £100,000 a year. The scheme has also made a contribution towards the company's successful ISO 14001 accreditation. More significant perhaps is the fact that the trust shown between the unions and management has increased and that environmental issues are now very much on the workforce agenda.

This concept of a workforce-led approach to saving energy and reducing waste in the workplace or, as the EEBPP calls it, a Partnership for Best Practice, was pioneered in the UK. Indications are that this country actually leads the rest of Europe in this type of initiative.

Improvement through involvement

Energy savings can be, and are, made through improvements in technology and working practices. However, without the support of the entire workforce, schemes are much less likely to be effective. Perkins Engines has shown that, with the enthusiastic support of an environmentally aware union, the involvement of shopfloor workers and suitable educational material – in this case provided by the EEBPP – total employee involvement can be just as powerful a tool for sustainable development as the introduction of new technology. Above all, common sense and a willingness to make improvements are often more effective at reducing energy consumption and minimising waste than making changes to manufacturing processes. By beginning with no cost, or low-cost improvements, significant savings can be made quickly.

Workforce interest and the EEBPP

The Trade Unions and Sustainable Development Advisory Committee (TUSDAC) was established in 1998 to mirror the activities of ACBE, the Advisory Committee for Business and the Environment. With over 60 per cent of UK union membership (20 per cent of the country's total workforce) represented by TUSDAC, the whole concept of energy efficiency and climate change is now being debated more widely than ever before at corporate level. For the first time, environmental issues are being addressed, not merely by senior managers but also by shopfloor workers.

To assist in rolling out the sustainable development message, the EEBPP has undertaken to develop examples of workforce-led environmental and energy best practice, so that union members can adopt these for themselves. A learning toolkit framework, to assist with worker education in this area, is also being developed by the programme.

Dr Uly Ma was trained as a metallurgist and worked in metals development, fuel gases and marketing. He is currently the Project Manager responsible for training and development in energy efficiency activities in the EEBPP. Uly previously led the development of the National Standards for Managing Energy and N/SVQs in Managing Energy.

The **Energy Efficiency Best Practice Programme** promotes energy efficiency through free, impartial, authoritative information and advice on energy efficiency techniques and technologies in industry, buildings and transport. This is provided through publications, videos, software, seminars, workshops and site visits. The programme also offers site-specific advice to energy users and design advice to building professionals. The programme is owned by the Department of the Environment, Transport and the Regions, and is run in partnership with the devolved administrations in Scotland, Wales and Northern Ireland.

Further information on the EEBPP, details of free publications and training material are available by calling the Environment and Energy Helpline: 0800 585794 or via the programme's web site: www.energy-efficiency.gov.uk

Perkins Engines manufactures diesel engines for a wide range of industrial, commercial, military and marine applications. The company, now part of the Caterpillar Group, employs 3,800 people at its Peterborough site. It is committed to improving the workplace and reducing the environmental impact of its operations, and has strong community links. Its commitment is demonstrated through its becoming a 'Making a Corporate Commitment' signatory, activities with Peterborough Environment City Trust, Peterborough Citywide Energy Audit and the National Schools programme.

27

The Challenges of Renewable Energy

Fred Dinning,
ScottishPower

This chapter describes how the ScottishPower group has established a formalised approach to environmental governance based on ISO 14001 and meeting the requirements of the Turnbull recommendations, but also making use of a practical approach to sustainable development. It describes how values and principles define strategic approach, how that is turned into corporate vision and how that vision is translated into key performance indicators and targets and, finally, how the process is audited and reported, both internally and externally.

Energy policy in the UK and Continental Europe today has three main sets of drivers: increased competition to bring down costs and make our industry more competitive; security of supply to keep the lights on at times of peak demand and to avoid overdependence on any one fuel; and finally protecting the environment from harmful emissions. Renewable energy offers considerable prospects as it can reduce our dependence on exhaustible fossil fuel reserves and offers emission-free energy. The UK is rich in a number of renewable energy resources, particularly wind (see Figure 27.1) and wave power. Feasibility studies undertaken on behalf of the government indicate that if all the technically feasible onshore wind reserves in Scotland were exploited they could meet the total energy needs of the UK. The shallow sections of the North Sea offer vast possibilities for offshore wind generation, and the coastline of our western shores are among the richest potential sources of wave energy. Waste to energy, biomass (burning cropped materials to recover energy) and solar energy either passively in building design or for direct generation of energy could have a role to play.

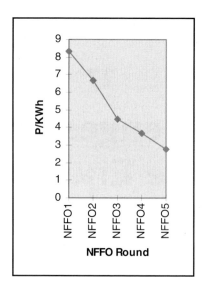

Figure 27.1 *Fall in NFFO bid prices*

So what barriers are preventing the rapid deployment of renewable energy in a liberalised market? The principle issue is cost. While renewable energy prices have tumbled, they still in the vast majority of instances fail to compete with conventional power generation from coal, nuclear and natural gas. Some argue that they should be given a guaranteed price premium, as they are in certain parts of Europe. Others argue that if the cost of environmental damage were included in the cost of fossil-fired generation (the externality cost), then renewables would be cost competitive and hence seek such mechanisms as a carbon tax to balance up the economics in favour of renewables.

These mechanisms are in stark contrast to the UK's renewable support mechanism, the Non Fossil Fuel Obligation (NFFO) in England and Wales, the Scottish Renewables Obligation (SRO) and the Northern Ireland Renewables Obligation (NIRO). The UK obligations set out to deliver set amounts of new, renewable generation from a fixed range of technologies on a cost-competitive basis. With no guaranteed price premium and a fixed amount of demand to be met costs have tumbled so that some technologies (such as onshore wind and landfill gas) are now close to competitiveness. Nor has the planning system helped. To secure the most cost-effective sites means selection of the most resource-rich renewable energy locations. In the case of wind this has led to some developers coming into conflict with communities and wildlife preservation groups. Since bids do not depend on planning consent having been secured many projects, having secured an NFFO contract, then fail at that hurdle and are never built. Completion rates have fallen to under 30 per cent for later NFFO bidding rounds.

There is also the issue of electricity system interconnection. Many developers argue that the cost of connection and use of the electricity system should be free (as it is in some parts of Europe). Additionally they argue that having generation in remote areas avoids the need to strengthen the network and reduces network losses by reversing the power flow. Sadly there is no 'free lunch'. The wind does not always blow, and customers want the lights to stay on. The electricity network is like the roots and branches of a tree. It was designed to carry bulk power from the strongly interconnected centres of fossil-fired power stations and bulk load points cities. Rural areas were supplied by more weakly connected lines of lower carrying capacity. The result of connecting power at these weak points can be voltage fluctuation and system instability. As the UK Regulation system requires fair and equal access to all, with cost-reflective charging, the renewable generator is required to bear the costs. These costs, plus the additional ones of the renewable energy, are passed on to all customers via the NFFO.

But NFFO and its counterparts have not been the only way forward. Many studies suggest that a marketplace for 'green energy' exists and that up to 10 per cent of customers might sign for energy from renewable sources even at a price premium. A number of energy suppliers, keen to develop their renewable resources more rapidly than waiting for NFFO rounds would allow, have developed 'green tariffs' and begun to sign up customers.

So has the UK system failed, and what characteristics are needed to succeed as a renewable developer? From the standpoint of delivering large quantities of renewables the UK system has clearly failed, whereas the Danish and German systems have succeeded. However the UK system has very effectively driven costs down and the UK customer gets his or her renewable energy for much less than his or her European counterpart. Arguably renewable developers have had to become much more sophisticated. In becoming one of the major renewable developers in the UK we in ScottishPower have had to become very astute in securing the best commercial deals for plant and equipment. We have had to build and maintain strong relationships with the main wildlife conservation groups and local communities. For instance we have recently secured planning consent for what will be the most powerful wind farm in the UK, at Beinn a Turic in Kintyre. We worked very closely with groups such as Scottish Natural Heritage and the RSPB in carrying out site assessment and will spend over £2 million on habitat creation measures to protect two golden eagles that hunt in the area of the wind farm. We have also sought to ensure that local employment is created in the construction of the wind farm. Setting ourselves a target of 10 per cent of our energy supplied by 2010 in addition to NFFO, SRO and NIRO projects we have launched a green tariff across our UK-wide customer base. The UK regime has required developers such as ourselves to develop a whole range of commercial, environmental and customer-focused skills that will serve us all well as the market develops.

So what of the future? The UK has recently consulted on how the aim of 10 per cent of energy secured from renewable resources by 2010 can be achieved. The sale of 'green energy' is likely to be encouraged, using certification schemes such as that recently developed for the DTI by the Energy Savings Trust, but customers will have to be found and retained. It seems probable that all energy suppliers operating in the new competitive energy markets will have to secure a portion of their energy from renewable resources, but they will need to keep costs down. The planning system may well be streamlined, but public acceptance will still require hard work. While there may well be more and larger NFFO rounds it seems unlikely that the competitive element will be taken away, as it has proven a sure way to control costs. Reconfiguring the electricity network to take large amounts of renewables will not be cheap, so developers who can work to control their impact or develop resource in better connected areas will fare better. Future market prospects are bright, but there will be no free lunches in balancing the needs of competition, security and the environment.

R J (Fred) Dinning is responsible for keeping the ScottishPower Board fully briefed on environmental issues, setting the strategic direction for the group's environmental thinking and ensuring that issue management and governance systems are in place and effective. He oversees the production of the Company Environment Report and organises and leads annual reviews of environmental risk management and governance and in the environmental aspects of Due Diligence during acquisitions. He also writes regular articles on the moral and ethical dilemmas society faces in the development and adoption of new technologies.

ScottishPower is the UK's leading multi-utility, and one of the top 10 utilities globally. It serves over 6 million customers in the UK and the US Pacific Northwest, with a turnover in excess of £5 billion. It has won numerous environmental awards, ranking as top utility in the FT/Business in the Environment survey and runner-up in the 2000 ACCA Awards.

ScottishPower
Cathcart Business Park
Spean Street
Glasgow G44 4BE

website: www.scottishpower.co.uk

28

The Future is Bright, the Future is Blue

Alison Hill,
British Wind Energy Association

Wind has been the fastest-growing energy sector worldwide for four years running (see Figure 28.1). One year ago the global wind energy industry passed a landmark 8,000MW of installed capacity; this figure now stands at 12,000MW – an increase of 50 per cent. This capacity is projected to double within two years, a rate of growth even faster than the most optimistic of industry projections.

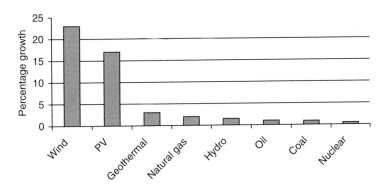

Figure 28.1 *Global growth by energy source, annual average, 1990–1998*

Source: REPP, Worldwatch 1999

Europe has continued to lead the way in developing wind energy, both in terms of generating capacity installed and export of turbines and associated expertise. The potential in Britain has yet to be fully realised. With the highest

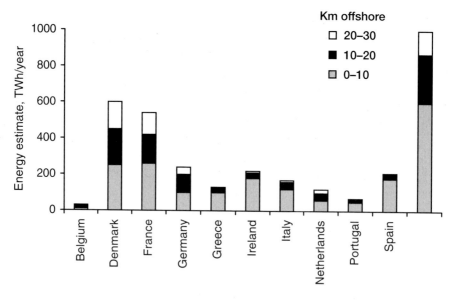

Figure 28.2 *European offshore resources*

wind resource in Europe, both at sea and onshore (see Figure 28.2), the British Wind Energy Association (BWEA) firmly believes that Britain is poised on the brink of a new energy epoch.

Part of the global expansion comes from advances in the technology itself. Wind turbine prices have fallen by a factor of at least three during the last decade, with the annual energy output per turbine increasing 100-fold. Increasingly energy efficient machines have more than halved the cost of electricity generated from turbines (see Figure 28.3). At good wind sites it is

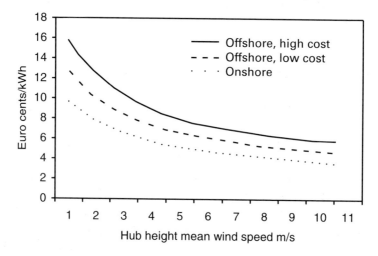

Figure 28.3 *Electricity costs*

already fully competitive with conventional power sources such as nuclear and fossil fuels. These costs will continue to fall as technology improves. The other part must surely lie with the increasing realisation of the need to reduce global carbon dioxide emissions.

In the UK, concomitant with figures on greenhouse gas reduction agreed under the Kyoto Protocol, the government has set a target of 10 per cent of all electricity generation by 2010 to come from renewable technologies. BWEA calculations show that 6 per cent of the forecast demand could be met by wind alone, from the existing industry base. A higher level could be met by an expanded industry, likely to emerge in support of more ambitious government programmes. Wind is the most advanced and commercially available of renewable energy technologies, and it is also among the cheapest.

Conservative estimates, however, suggest that wind is capable of supplying 10 per cent of the world's electricity needs by 2020, even if global electricity use doubles as predicted. In their joint report, *Wind Force10*, the Forum for Renewable Energy, Greenpeace International and the European Wind Energy Association provide a practical blueprint showing how this can be done. Achieving this goal would create 1.7 million jobs worldwide with annual investments increasing from £3.5 billion to almost £50 billion in 2020 and an annual reduction of 1.78 billion metric tons of carbon dioxide.

Meanwhile in Britain the day-to-day business of the BWEA continues to be the furtherance of its members' interests, and therefore the wind energy industry. 'We offer our members the latest information, the best intelligence and the highest possible level of access to the system,' says Chief Executive Nick Goodall. 'These benefits are clear to businesses looking to enter the industry, as well as those already there.' The BWEA carries out activities not only through the conventional 'fact-finding and lobbying' route, but also through the initiation of stakeholder dialogues. Several working groups have been established to examine key areas in the progress of the wind energy industry in the UK. Offshore, planning and now regional targets have all been addressed in the resulting dialogues, which bring together all those with an interest in the same place at the same time. Issues of concern can then be identified and addressed.

As a result the industry is now ready to take the next step in its development. Of all the renewable energy technologies, offshore wind energy has possibly the most favourable combination of the key attributes of resource, energy cost and risk. The Renewable Energy Advisory Group has calculated that the onshore resource alone was greater than the entire electricity demand of the UK in 1992; the offshore resource is even greater still.

Wind speeds at sea are higher and more constant, giving a higher initial output, and the energy efficiency overall can be greatly increased by increasing the size of the turbine. There is a strong economic case for building large offshore machines because the foundation and cabling costs are largely independent of the size of the machine. Energy costs will initially be higher than for onshore installations, roughly 40 per cent, but it is hoped that this could be addressed by a government support scheme, similar to the NFFO mechanism currently in place onshore.

The world's wind resources are unlikely ever to be a limiting factor in the utilisation of wind power for electricity production. Even with wind power generating 10 per cent of the world's electricity by 2020, this still leaves most of the resource untapped. In Britain the only limiting factor may prove to be the willingness of the government to set in place a coherent energy policy, supportive of renewable energies. The proposed electricity trading arrangements, together with the lack of a future support mechanism may mean that the government will be unable to meet its own targets.

The UK, with its offshore expertise and huge wind resource, is well placed to benefit from this rapid growth industry. There are tremendous opportunities for job creation in the fields of offshore services, construction and, of course, in the manufacture of wind turbines and their components, together with the continued development of the onshore resource. However, the full potential for the UK will only be realised if the government promotes plans that enable the economic and environmental advantages of wind farms to be exploited, and does so soon.

Alison Hill is Communications Manager for the British Wind Energy Association. Her background in both science and media led to an almost inevitable move into the renewable energy sector, and she sees her future firmly linked to wind power. 'Knowing you're making a difference, doing what you're good at, makes for a more positive attitude all round.'

The British Wind Energy Association
26 Spring Street
London W2 1JA

tel: 020 7402 7102
fax: 020 7402 7107
e-mail: info@bwea.com
website: www.bwea.com

From its early beginnings more than 20 years ago as the professional association for researchers and enthusiasts in the then embryonic wind industry, the **British Wind Energy Association** has developed into the largest renewable energy trade association in the UK. With a membership of over 500, including more than 100 corporate members, generating an annual turnover of 528,000 ecus, the BWEA is uniquely placed to consolidate and extend the wind energy industry in the UK.

Section VIII

Pollution Prevention, Air, Water and IPPC

29

European Air Quality Directives: Which Way is the Wind Blowing?

Charlotte Granville-West,
Confederation of British Industry

The local effects of air quality have long been recognised – for example, the UK Alkali Act was established in 1863 to improve conditions adjacent to installations; but it has only been in more recent times that a transboundary approach to air quality has emerged. This was formally recognised in 1979 through the Geneva Convention for the Control of Long-Range Transboundary Air Pollution (CLRTAP), which set general principles for international co-operation on air quality.

The European approach

As well as the EU being a signatory to a variety of such international treaties, a number of European initiatives have emerged in parallel, particularly in the wake of concerns over ozone, acidification and, more recently, climate change. The major focus of policy to improve air quality has been the reduction of emissions from all sources, in particular photochemical substances and those leading to acidification. According to the European Environment Agency these measures have had varying degrees of success, resulting in a general decline of air pollutants, with targets for 2000 (set in the fifth environmental action programme) being reached for sulphur dioxide.[1] The raft of measures introduced

[1] *Environmental Signals 2000*, European Environment Agency.

so far, known as the 'Reference Scenario', are expected to protect almost all EU ecosystems from acidification and deliver the health-based ozone target of an eight-hour average of 160 µg/m³ (80 parts per billion) proposed by the US Environmental Protection Agency.

Historically, air quality issues have tended to be addressed individually, resulting in a number of initiatives to address ozone depletion or acidification, but no overall strategy or co-ordination in regulation. A recent attempt was made to improve the situation through the creation of the Directive on ambient air quality assessment and management (96/62/EC), which is intended to be a framework bringing together quality standards for individual pollutants through the creation of Daughter Directives. Initial proposals include sulphur dioxide, nitrogen dioxide, fine particulate matter, ozone and some heavy metals, but as new scientific evidence becomes available it is likely that the list will grow, for example dioxins are already being considered. Despite the framework Directive there remains a dual approach within the EU, the use of *quality standards for Member States*, as well as targeting *emissions from specific sources* through separate initiatives such as the IPPC Directive (96/61/EC) and the Large Combustion Plant Directive (88/609/EC). This has resulted in several lists of the most important pollutants established and controlled under different Directives. It is not yet clear what effect the push for integration from the IPPC Directive will have on air quality, but clearly a more coherent approach would be of benefit to business.

Since the Amsterdam Treaty came into effect on 1 May 1999 the European Parliament's powers have been enhanced through an extension of Co-decision Procedure to cover environmental legislation. Current air quality proposals before Parliament are therefore subject to this decision-making process, which gives the Parliament the right of joint decision making with the Council of Ministers. The effect of this has been to slow down decision making, as a common position has to be reached and agreed upon. Ultimately the number of environmental Directives going to Conciliation is expected to increase – air quality being no exception.

Current proposals

Proposals for new air quality Directives continue the issue-based approach and currently include Daughter Directives to 96/62/EC, a review of the Large Combustion Plant Directive and a new National Emissions Ceilings Directive (NECD). The NECD seeks to reduce acidification by setting national emission limits for the same four pollutants as in the recent multi-pollutant protocol to CLRTAP: sulphur dioxide, nitrogen oxides, volatile organic compounds and ammonia. This is on top of the so-called 'Reference Scenario', with Commission proposals projected to cost a further 7.5 billion euros per annum.[2] On 22 June 2000 a common position was reached between Member States that proposed limits between the Commission proposal and the multi-pollutant protocol, but still representing a significant commitment, requiring large cuts and subsequent costs to industry. Unlike other environmental protection measures the limits proposed by the Commission for the NECD suffer from being based on predictive modelling which, as with all models, is subject to significant uncertainty. This raises the question of how such models should be used in setting strict standards for air quality. Overreliance on predicted trends without a proper review of the actual effect of existing measures can lead to inefficient and costly regulation.

Future trends

Air quality will continue to be a priority for the EU both during the accession process and through the sixth Action programme, which is likely to include binding targets. Accession candidates will be expected to focus resources on the implementation of the air framework directive (96/62/EC), one of the most expensive.[3] Since the arrival of the new Commissioner Wallstrom to DG ENV it has been stated that review of existing legislation, its implementation and effects would be increasingly important in assessing the need for new Directives.[4] Coupled with a more integrated organisational approach and less reliance on predicted trends this would be a very positive step towards a more efficient, coherent and successful policy for the EU.

Charlotte Granville-West is a senior policy adviser, Business Environment, at the CBI. She has been working with the CBI for a year on a wide range of issues relating to better regulation. This has included air quality, water issues, IPPC, utilities review and the Environment Agency.

---

[2] Commission proposal, COM(1999)125final, Annex I, Table 8, p 47.
[3] 'Enlarging the Environment' (1999) Newsletter from the European Commission on Environmental Approximation, (DG ENV), no.15, August.
[4] Wallstrom M (2000) European Commissioner for Environment *European environmental policy: the next steps*, ERM/Green Alliance Forum, London, 28 February.

CBI
Centre Point
103 New Oxford Street
London WC1A 1DU

tel: 020 7395 8053
e-mail: charlotte.granville-west@cbi.org.uk

The work of the **CBI Environment Group** covers all aspect of environment, health and safety and their relationship to business. Listed below are some of the key policy areas the Group is currently engaged in:

- **Climate change** – climate change levy, emissions trading
- **Producer responsibility** – environmental liability, waste from electrical and electronic equipment – proposed Directive
- **Better regulation** – IPPC, Environment Agency
- **Risk management** – environmental reporting
- **Health and safety** – corporate killing, revitalising health and safety.

30

Implications for Industry of the EU Water Framework Directive

David Taylor,
AstraZeneca

The European Community has been developing its policy on water since the 1970s, but some of the early Directives were only agreed after considerable compromises and there are many gaps, overlaps and inconsistencies in the current policy. The Treaty of European Union (1992), signed at Maastricht, dramatically changed the position by establishing environmental protection as part of the treaty obligations and by making these issues subject to qualified majority voting. In addition, the Community now contains 15 states and further enlargement is imminent. As a result the Commission has undertaken a comprehensive review of water policy, resulting in the Proposed Water Framework Directive that is intended to replace most of the early Directives.

Following several years of detailed negotiation, and a hectic final round of conciliation talks between the Council, Commission and Parliament, the final form of the proposal and its major implications have now become clear. There are still formal procedures to go through before the proposal is finally approved, but no significant changes are now expected.

Most of the underlying concepts on which the Directive is based will be familiar to UK readers: water resources, both quantity and quality, are to be managed on a river basin basis; each area must have an improvement plan that is agreed with the local stakeholders; users must pay the economic cost of their use of the water cycle; waste discharges must all be licensed and are to be controlled by a combination of emission standards and environmental quality objectives; there is to be a list of priority substances that need special attention; coloured maps are to be produced periodically to demonstrate the rate of improvement.

However, despite the process being familiar, there are major implications for UK business. These relate to the improvement targets and timetables specified and fall into three areas:

- **Surface water** – All EU surface waters will have to be improved to 'good quality' by 2016. Surface waters include estuaries and inshore coastal waters together with rivers, streams and lakes, although there are exclusions for very small catchments and small ponds.

  Quality will be measured with respect to meeting all chemical quality standards together with achieving good ecological standards, compared to unpolluted reference waters. Some derogations are allowed, where improvement is 'infeasible', or where the natural environment has been 'heavily modified' for human benefit, but these will be tightly constrained. The objective is to restore and protect the quality of all EU surface waters without the necessity for removing all dams, ports and harbours.

- **Groundwater** – All groundwaters will also need to be of 'good quality' by 2016. This means protect, enhance and restore the quality of all significant aquifers. The practicality of this objective and the interpretation of what 'good quality' means are more difficult to define than for surface waters and in practice the debate is still proceeding as to what this actually means. This will be resolved in the next two to three years in a subsequent 'Daughter' Directive.

- **Hazardous substances** – There will be a requirement to severely restrict and in some cases to eliminate the discharge of hazardous substances. A priority list will be established by the Commission, which will put forward uniform emission standards and environmental quality objectives that must be implemented across the whole community. A set of 'priority hazardous substances' will also be identified, whose discharge to the aquatic environment must be reduced to zero within 20 years of their identification. The materials concerned are not yet selected, and negotiation continues in this area.

The Water Framework Directive is a very large and complex legal instrument and it will take time, and the development of case law, to fully identify its consequences. The majority of its requirements actually mimic current UK practice and thus will not impose additional requirements on business. However, the 2016 deadline for the achievement of good quality surface waters may lead to some acceleration of current Environment Agency (EA) programmes, particularly in areas of poor water quality, and companies need to be prepared for this and for significant problems and costs arising from the groundwater and hazardous substances sections.

Significant restrictions can be expected on groundwater abstraction in areas where water resources are poor, and potentially very significant technical problems and costs are likely to be incurred by companies that are required to restore contaminated aquifers. In particular, the identification of responsibility will raise many issues. It would be in the interests of companies to ascertain the quality of the groundwater beneath their installations and whether any contamination present might be ascribed to them.

Controls at Community level on the substances on the priority list may also involve some additional costs, probably not dissimilar to those that might have been expected under planned UK EA regulations. However, the requirement to totally eliminate the discharge of the 'priority hazardous substances' may have profound effects on business, depending on which substances are selected. It is likely that the only way to achieve the objective of 'zero discharge' for substances in this category will be to cease their manufacture and use.

The consequences that stem from this 'zero emission' policy, in terms of the products that will no longer be available, have not yet been appreciated by either politicians, regulators or the population at large. Stakeholders currently have the expectation that any substance that is 'banned' will instantly be replaced by a 'safer' alternative at the same price. It will be in the interests of business to bring home to all stakeholders, at the earliest possible opportunity, the consequences that will flow from this section of the Directive.

Finally, the monitoring requirements of this Directive will require a very large increase in activity by the UK agencies. It would be naive to assume that this additional cost will not find its way back to industry.

Professor David Taylor is Environmental Foresight Manager in the Corporate SHE Department of AstraZeneca. He is a chartered chemist and Fellow of both the Royal Society of Chemistry and the Institute of Water and Environmental Management. In 1998 he was appointed Visiting Professor in the Department of Earth Sciences at the University of Liverpool. Dr Taylor currently chairs the CEFIC Environment Core Group, the Royal Society of Chemistry's Environment, Health and Safety Committee, and the Management Board of the Green Chemistry Network, and is a member of the CBI Environmental Affairs Committee.

Professor David Taylor
Corporate SHE Department
Brixham Environmental Laboratory
Brixham
Devon
TQ5 8BA

tel: 01803 882882
fax: 01803 882974
e-mail: david.taylor@brixham.astrazeneca.com
website: http://www.brixham.astrazeneca.com/

AstraZeneca is one of the world's leading pharmaceutical companies. It is a strongly research-driven organisation with a formidable range of products designed to fight disease in important areas of medical need. The company was formed in April 1999 by the merger of Astra AB and Zeneca Group plc.

31

Practical Steps for IPPC Implementation

Steve Bradley,
British Sugar

The Integrated Pollution Prevention and Control (IPPC) Directive places duties on those installations (factories) that come under it to:

- take all appropriate preventive measures against pollution;
- prevent significant pollution;
- avoid waste production at source;
- recover waste or, where that is not reasonably practicable, dispose of waste in such a way as to minimise its environmental impact;
- use energy efficiently;
- prevent accidents and limit the consequences of those that do occur;
- avoid land contamination.

The Directive will be brought into UK law under the Pollution Prevention and Control (PPC) Regulations 2000. The first step is to determine whether the regulations apply to your operations; schedule 1 (not reproduced here) provides the full list. The scope of the installations covered is defined sometimes by a tonnage capacity, eg *more than 75 tonnes per day capacity*. The word 'capacity' means historical throughput rather than a theoretical maximum.

An installation does not necessarily mean the same as site or factory. It is important to define where on the site the Regulations apply. If there are multiple occupiers on the same site then it is essential to define clearly what is covered by the PPC Regulations and who has responsibility for what. This must be documented.

All new installations subject to the PPC Regulations must comply with the requirements immediately. For existing installations there will be a phased implementation such that the workload for the regulators will be spread across many years and not bunched together.

It is most strongly recommended that companies set up a team to manage the whole project. The team needs to be visibly supported by the company so that it has the necessary influence within the organisation to ensure that work is done at the correct priority. The team might usefully have an environmental specialist, a process engineer, a control engineer, a production manager and the relevant operators. Clearly people with different skills need to be involved at different times as appropriate.

A key action is to quantify exactly what the installation will consume and produce. This production process should be analysed with respect to:

- units of consumption of raw materials per tonne of product (or some other equally useful denominator);
- the amount energy consumed per tonne of product;
- units of emissions to air and water per tonne of product;
- units of solid and packaging waste per tonne of product;
- the quantity and type of noise and vibration produced;
- the likely absolute tonnages of consumption and emissions per year;
- the impacts on the local environmental conditions.

This straightforward process balance can often yield some surprises and provide pointers for action.

The main duty is for permitted installations to make sure that BAT is used to prevent or to minimise pollution and to protect the environment as a whole. The European IPPC Bureau in Seville is drawing up BAT Reference Documents (BREFs) to provide guidance. It is worth looking at their Internet site: http://eippcb.jrc.es/ to see which will apply and to check progress. It is very important to compare the BREFs with your operations to help determine BAT.

One important feature of all PPC permits is that of monitoring and measuring. The consumption of raw materials and the emissions to air, land and water will need to be quantified. All monitoring data will need to be reported to the regulator and be placed on the public register and the Internet. Operators will therefore need reliable systems. Operators are advised to propose to the regulator what the monitoring regime will be, as this provides a higher level of comfort to the regulator and puts the operator in some form of control.

Management systems will need to be set up to warn when certain statutory events are due, for example when the next emissions monitoring event should take place. It is too easy to leave these issues to people's memory and for them be forgotten.

Of increasing importance is the effect of industrial processes on ecological systems. In areas of wetlands there will be considerable sensitivity regarding water abstraction. In other areas of the country there could be great sensitivity to emissions of certain compounds to air. Overall the concept of biodiversity is of great importance and one that is relatively new to industry in general. The regulators will be required to consult English Nature, the Ministry of Agriculture, Fisheries and Food and other interested parties to gain their opinion. Therefore operators of permitted installations will need to have carried out ecological risk assessments for their operations. All of this information will reside in the public register.

Linked to this is the relatively new requirement to prevent pollution to groundwaters. If it is a relevant topic then the site operators can expect to have to carry out hydrogeological surveys and to determine whether the soils provide sufficient purification to prevent significant pollution. This will undoubtedly be a major cost and will take a long time to determine.

An explicit requirement of the PPC Regulations is the need to avoid pollution when finally closing the installation and to return the site to a satisfactory state. The obvious question is: to what standard will the land need to be cleaned? No easy answers are provided here – monitor government guidance.

The other novel feature is the duty to take the necessary measures to prevent accidents and to limit their consequences. This applies to both environmental accidents and to human safety and health. Question: are you able to prove that you have adequate control and that you routinely rehearse emergency procedures?

Allow sufficient time for the necessary administrative controls and public consultation when applying for a permit. It is most advisable to 'warm up' the regulator well in advance. This early consultation will allow you to understand the information needs and fill any potential gaps in advance. For large installations it could take well in excess of a year to provide the regulator with staged applications and numerous drafts so that both sides are well informed and without surprises. Regulators always appreciate consultation to allow them to influence where appropriate at the stage at which changes are least costly to the business.

It is very clear that the public can be affected by emissions and operations and so has a right to understand what is taking place in industry. Politicians take a keen interest in such matters and are very able to make life difficult for organisations that do not appear to have the proper regard for neighbours. In short, industry is now firmly accountable for its actions to the public and the public expects and demands high standards. Can you handle this in-house or is external resource required? The costs will be significant per installation, but the costs of not being prepared will be far higher!

Steve Bradley is Head of Safety, Health and Environment for British Sugar. He has been involved in the development of the IPPC Directive at a European level and with its practical implementation within the UK. He is the chairman of the Environment Working Group of the European sugar industry and is an active member of the UK Food and Drink Federation's IPPC Group.

e-mail: sbradley@britishsugar.co.uk

British Sugar is the sole processor of the UK sugar beet crop, grown by some 8,500 farmers on 150,000 hectares, mainly in the eastern counties. The processing season, known as the 'campaign', usually lasts from September to the end of February. The British beet sugar industry supports some 23,000 jobs in the farming, processing and transport industries. Producing some 1.4 million tonnes of sugar, British Sugar is the leading supplier to the UK providing more than half the country's sugar requirements.

British Sugar plc
Central Offices
Oundle Road
Peterborough PE2 9QU

tel: 01733 563 171
website: www.britishsugar.co.uk

32

The Regulator's Perspective on IPPC

Stuart Stearn,
Environment Agency

A new system of pollution control. Another batch of Regulations. Another Directive. A hard-pressed industrialist may well be forgiven for thinking that he does not need this. However I hope you will find as you look through this book that there is much in IPPC that you do already and that the principals on which it is based are consistent with good business management principals.

Integrated Pollution Prevention Control (IPPC) applies an integrated environmental approach to the regulation of certain industrial activities. This means that emissions to air and water (including discharges to sewer and land) plus a range of other environmental effects, must be considered together. It also means that regulators must set permit conditions so as to achieve a high level of protection for the environment as a whole. These conditions are firmly based on the concept of using the best available techniques, which balances the cost to the operator against the benefits to the environment. IPPC aims to prevent the emissions and waste production and where this is not practical to reduce them to acceptable levels. IPPC also takes the integrated approach from the initial permit through to the restoration of sites when industrial activities cease.

IPPC is based on the idea that operators have responsibilities for what they do. Operators know what to produce to satisfy the market. They need to be responsible for the consequences of that and produce their products with the minimum disruption to others. Therefore installations that carry out the various prescribed activities must be operated to ensure that:

- best available techniques are used to prevent pollution;
- no significant pollution is caused;

- waste is prevented, minimised or recovered for safe disposal;
- energy is used efficiently;
- accidents are prevented and their consequences limited;
- the site is returned to a satisfactory state when operations cease.

Although this sounds a long list all operators should be doing this anyway. It is not just about the control of pollution, it is about saving waste and saving energy, and all shareholders should be regularly questioning their directors about this. Even the issue of site restoration, which has troubled some, is a shareholder interest. IPPC is in no way retrospective; it will not require operators to clear pre-existing pollution, but it is reasonable for a shareholder to make sure that the asset value of the company is not eroded any further by allowing land to become more contaminated, and accident prevention is just good sense as a policy in reducing exposure to long-term legal risks.

So the process, familiar already to those regulated under IPC, is based on three questions. The first is what are you doing, the second why you are doing it this way and the third is what are the consequences for the environment. This last question matters because since one person's release is another person's pollution there are legal requirements that environmental quality standards are not breached. These are all questions that operators should be asking themselves as a matter of course.

With new regulations naturally come new definitions, new ways of doing things, and this inevitably causes some uncertainty about how to proceed. For example, the concept of 'installation' is defined in the Regulations and there is advice in the DETR's 'Practical Guide', but there will need to be careful consideration in particular cases to make sure the right units are picked out to be the focus of regulation. The Environment Agency is certainly very conscious of this point. There are issues, such as energy efficiency, that are new and where there are interactions with the emerging regime of climate change levy. There are new interactions between the Environment Agency and local authorities in respect of the regulation of noise that will require sensitive and sensible handling to ensure a successful outcome. But the system remains an opportunity for industry at large to show what it can do to put the environment at the heart of its thinking, to see it as another factor to consider when developing its future plans, including finance and employee liability, in order to develop effective and economically sustainable programmes for steadily improving its environmental performances in a competitive environment.

The Environment Agency has completed its consultation on its regulatory package containing application forms and guidance for those wishing to apply for permits. We have been very grateful for the time and trouble people have taken to offer us comments, which we will certainly reflect. We welcome the CBI's Handbook as a further resource for operators to use to help them through the new regime.

IPPC is equally having its effect on the Environment Agency. We have had to work hard to produce what we believe is a closely integrated regulatory package. We have also had to think very hard about the procedures we need to generate for internal use to ensure that we operate consistently across the breadth of the Agency's interests. We believe that this has the potential to improve the way we do things. It can also provide a framework to help industry come to structured decisions where the environment is concerned in a way that highlights the interaction between business performance and the environment.

This interaction is often set out as being less pollution equals more costs. We think that the working of the IPPC Directive will eventually show that this is not always the case.

IPPC does not focus on the eventual emission limits allowed. It focuses instead on the decision process which eventually leads to that and the Agency's aim is to ask searching questions of industry so that we can be fully satisfied that that process has been sound, thorough and rigorous. We believe the CBI's Handbook will complement our own efforts to make sure that this happens.

Dr Stuart Stearn read engineering at Cambridge before joining the Central Electricity Generating Board where he worked on a range of engineering problems associated with nuclear and conventional generating and transmission equipment. He then worked for the Department of the Environment's radio-chemical inspectorate. In 1989, he became head of the North West Regional Office of Her Majesty's Inspectorate of Pollution (HMIP) and in 1992 began managing the Inspectorate's research and development programme and policy development on radioactive waste management and integrated pollution control. In 1996, HMIP became part of the Environment Agency where Dr Stearn is currently Principal Technical Co-ordinator on the IPPC Project Team

Environment Agency
Steel House
11 Tothill Street
LONDON
SW1H 9NF

The **Environment Agency** for England and Wales was established by the Environment Act 1995 and became fully operational on 1 April 1996. It regulates many activities and processes that impact on the environment, including industrial and business activities that cause, or have the potential to cause, pollution to air, land and water. The Agency is the regulator for A(1) activities under the PPC Regulations. At the same time, it must have regard to the conservation of features of special interest. It also regulates the abstraction and storage of water, the exploitation of freshwater fisheries and the use of certain river navigations. As an operator the Agency has a

33

Towards Sustainable Manufacturing: Technology Trends and Drivers

Mark Hilton,
ECOTEC Research & Consulting

Until the mid 1970s we lived in a world where resource depletion and pollution were seen to be of limited importance. The following decades, however, brought an increasing range of EU and UK pollution control regulation and with it an era of end of pipe 'environmental technologies' to allow compliance. In the late 1980s and the 1990s there was a shift towards more preventive regulation and an increasing awareness of 'eco-efficiency' concepts such as waste minimisation and energy efficiency. This coincided with a move towards improved quality management and hence companies began to look for technologies that brought environmental and/or workplace benefits along with quality and efficiency improvements. Businesses therefore began adopting systems that:

- produced better quality products;
- maximised material yield and minimised losses;
- used less energy and water;
- took up less space on the shopfloor;
- were simpler and cheaper to operate and maintain;
- were inherently cleaner, producing less pollution and improving shopfloor conditions.

These 'cleaner technologies' therefore encompass all sorts of 'regular' process technologies and materials and the more proactive companies that have

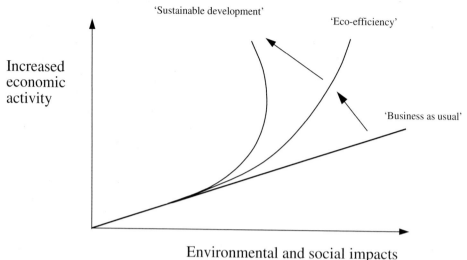

Figure 33.1 *Towards a sustainable growth path*

adopted them have moved on to a less damaging 'eco-efficiency' growth path (see Figure 33.1).

The 'end of pipe' technologies, while originally aimed at cleaning up emissions prior to discharge, have begun to play an increasingly important role in terms of resource recovery. Important technologies in this regard include:

- settlement/centrifugation/flotation (to separate high/low density materials from water/air);
- membrane filtration (physical separation for a wide variety of materials);
- ion exchange (eg for recovery of metals and electrically charged coatings);
- evaporation/condensation/distillation (eg solvent recovery from waste coatings);
- adsorption/desorption (eg solvent capture and recovery from gas streams).

While all of these technologies have made a very significant impact, it is perhaps membrane systems that have been applied most widely; they are found in many sectors including chemicals, pharmaceuticals, textiles, metal finishing, ceramics, food and drink, paper/pulp and chipboard/fibreboard manufacture. The type of membrane varies from simple cloth-type microfilters, to concentrate such materials as latex, oils, paint pigments, etc, to sophisticated polymer 'nano' and reverse osmosis filters that can take out very tiny particles (down to less than 0.001 of a micron) including metal ions, dyes, artificial flavours and various organic molecules.

The advantage that membranes have over certain other separation technologies is that they can: (a) leave materials uncontaminated for direct and trouble-free reuse without further processing; and (b) leave a clean effluent that can also be reused. While such systems can be quite expensive (although their cost has fallen in recent years) they typically pay for themselves within months. Where water quality is paramount, membranes can be used as the final 'polishing stage' in hybrid systems, regular settlement and filtration being used to protect the membranes. It is clear that the range of applications will continue to grow as new materials and systems are developed and as prices drop.

While aimed at abatement rather than recovery, it is worth noting that considerable developments have also taken place in terms of the biological treatment of industrial pollution. Relevant technologies here include:

- enhanced biological wastewater treatment, using more effective aeration techniques;
- bioadsorption, for example using reeds for the removal of heavy metals from effluents;
- biofiltration of volatile organic compounds and odour-laden airstreams.

Such technologies can significantly reduce the environmental impact of operating pollution control systems as the 'bugs' and biomass do most of the work, generally without much need for energy inputs, chemical additives, material replacement or other forms of maintenance. Traditionally used in municipal wastewater treatment, such biosystems have become more widely used in industrial applications in recent years.

Moving 'upstream' to cleaner process technologies and materials, one can identify a number for which there is scope for far greater use. These include:

- computerised monitoring and control systems (for process optimisation)
- 'natural' and water-based adhesives, coatings and laminates;
- high efficiency (eg high volume low pressure (HVLP) and electrostatic) spray guns;
- zero-emission (enclosed) chemical mixers;
- high efficiency (eg dry) vacuum pumps;
- high efficiency/low maintenance 'hosepumps' for effluents and abrasive materials;
- computer-controlled machines/robots for forming and coating;
- systems using microwave or ultrasound-assisted processing;
- high pressure cleaning in place systems for vessel/booth washing, etc;
- line 'pigging' systems for the mechanical cleaning of pipes;
- high efficiency (eg condensing) and low-NOx boilers.

What these various technologies have in common is that they can significantly reduce resource use while often also bringing productivity and quality gains. To give some examples: one chemical formulator has been able to move to a zero-discharge production process, saving water and product, through the computer-controlled reuse of vessel wash waters. In the food and speciality chemicals sectors the use of 'pigs' (rubber bungs that can be forced through even small pipelines using compressed air) is allowing close to zero product loss while also reducing water wastage and effluent charges.

Waste minimisation and cleaner technology in the ceramics sector
The ceramics sector offers a few good examples of the ways in which cleaner technologies and approaches can be applied to reduce wastage and improve efficiency. These include:

- use of peristaltic 'hosepumps' to deliver slip and move effluents;
- automated 'granulate' pressing of 'flatware', reducing clay, water and energy use;
- microwave and airless drying to reduce product cracking and wastage;
- robotic spraying using HVLP guns to improve glaze application efficiencies;
- use of high pressure wash systems on glaze lines to reduce water use;
- use of 'pigs' to recover glaze and slip from pipelines;
- use of filter presses and membranes to recover clay and glaze materials from effluents;
- use of settlement chambers and filters to recover clay and glaze particles from airstreams.

While much more needs to be done to spread these and other technologies, new drivers are changing the context in which firms have to operate. The most obvious pressures here relate to producer responsibility and product 'take back' regulations (eg the Packaging, Waste Electrical and Electronic Equipment and End-of-Life Vehicles Directives), which are driving the eco-design movement. This not only affects the products but also the manufacturing and remanufacturing processes.

   In terms of packaging, for example, design for recyclability considerations are leading to the move away from multi-material packs (eg plastic and paper) to single material packs (eg corrugated board), with associated process implications. In the electronics sector the technology needs to be there to deal with new types of lead-free solder and to allow simple and preferably automated disassembly. In this way 'environmental' technologies become yet more indistinguishable from 'regular' process technologies.

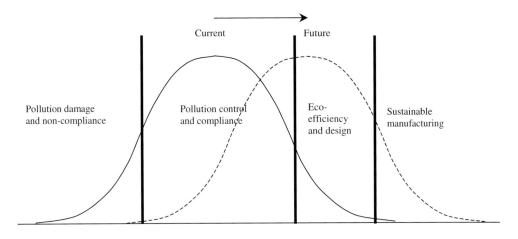

Figure 33.2 *The 'mainstreaming' of eco-concepts and technologies*

As Figure 33.2 indicates, eco-efficiency and eco-design ideas are slowly becoming more mainstream, a small progressive vanguard leading the compliant but cautious majority (with the 'resisters' trailing behind). More fundamental changes, however, are on the way. Considerable progress is being made, for example, in terms of manufacture using wastes and by-products (eg industrial ecology), renewable materials (eg biorefining) and 'green' chemistry. Example materials and technologies include:

- natural fibres (such as hemp and flax) for the manufacture of composite materials
- natural starches for the manufacture of packaging materials
- crop-derived and biodegradable plastics (eg polylactide, PLA, from maize)
- vegetable-derived chemicals to replace petrochemicals
- textile colouring without dyes (eg using chromophores).

Renewable energy (eg photovoltaics, fuel-cells, energy from waste) also has an increasingly important part to play, as will 'nano-technology' – tiny machines that will allow all kinds of process innovations. 'Cleaner technology' is therefore set to take us a step closer to the sustainable manufacturing paradigm, new materials and machines adding to the existing array of eco-efficient processes and products.

Mark Hilton is an environmental engineer (BSc, MSc) and senior consultant in ECOTEC's Environment Group specialising in industry good practice and the effects of policy and regulation. His particular fields of interest include:

- industrial eco-efficiency – waste minimisation, eco-design and sustainable production
- recycling and waste management strategy
- environmental support, education and training to small and medium-sized enterprises.

During his six years at ECOTEC he has worked for numerous private and public sector clients, the latter including the European Commission, the DETR, DTI, the Welsh Assembly, the Welsh Development Agency and the Local Enterprise Development Unit in Northern Ireland. He has also researched and written 20 'Good Practice Guides' and 'Case Studies' for the Environmental Technology Best Practice Programme. Prior to joining ECOTEC Mark worked for 10 years as an analyst in the advanced projects area of British Aerospace.

ECOTEC Research and Consulting Ltd
28–34 Albert Street
Birmingham B4 7UD

tel: 0121 616 3643
e-mail: mark_hilton@ecotec.co.uk

ECOTEC Research & Consulting is an independent, international, multidisciplinary consultancy organisation, founded at Aston University in 1983. The company now has around 140 full-time staff, based at its offices in Birmingham, London, Leeds, Brussels and Madrid. The staff come from a wide range of backgrounds and include specialists in engineering and technology, environmental sciences, economics, planning and business analysis. Almost all have a higher degree and experience working at a senior level in relevant parts of industry, commerce and the public sector. This breadth of expertise and experience allows the company to adopt an effective multi-disciplinary approach to defining, addressing and resolving problems on behalf of its clients.

34

Clean Technology Saves Money

Martin Gibson,
Environmental Technology Best
Practice Programme

Many companies are finding that by reducing their waste at source they not only improve their environmental performance but also save money. Waste minimisation, through no-cost or low-cost measures, has proven popular and can certainly lead to improvements. However, to make step changes in efficiency and cost savings, investment in cleaner technologies is usually required. By these I mean production technologies that are inherently more efficient and produce less waste than conventional technologies that treat pollution at the 'end of the pipe'.

The Environmental Technology Best Practice Programme (ETBPP) promotes the uptake of cleaner technologies because of their dual environmental and financial benefit to companies. It encourages companies to consider cleaner technologies when they are investing in new capital equipment. Unfortunately the purchase price of cleaner technology equipment is often higher than the price of more conventional technologies. However, cleaner technologies often generate more profit when the whole life of the equipment is considered. This is because operational savings from such technologies can far outweigh the higher initial investment costs.

The Programme has produced a number of case studies that show how companies have benefited from using cleaner technologies. These cover a range of industry sectors and technologies. Examples of four companies that have benefited from using cleaner technologies are covered here, but many more are available.

Questioning traditional approaches can sometimes turn a problem into a profit, as was demonstrated by cheese maker Joseph Heler Ltd. Conventional cheesemaking gives rise to whey, a watery waste containing potentially valuable proteins and sugars. As a waste, the whey needs to be disposed of carefully to avoid environmental damage. However, by investing in an integrated whey recovery system based on membrane technology, Joseph Heler Ltd was able to recover whey protein and lactose, and sell it. The company also produced clean water that could be reused.

The capital cost of installing the new technology was just over £1 million. However, the increased income from the sale of lactose and whey protein amounted to almost £700,000 per year. This gave a simple payback time for the equipment of about 18 months, after which, the savings contribute to increased profits. The ETBPP has produced a case study on what the company did entitled *Turning Waste Into Profit* (reference GC150).

Improved environmental awareness and the adoption of cleaner technology also led to cost and environmental savings for MacDermid Canning plc – a company that formulates surface finishing chemicals. An environmental review at its Albion Works highlighted several issues, in particular excessive water use, and provided the stimulus for action.

The company needed to install additional equipment to increase production capacity, and the environmental team presented a case for a cleaner technology option. Rather than simply expanding, by adding to what they had, the team's suggestion was a fully computerised mixing facility that was more energy efficient and had virtually no discharge, with washwaters being reused in later batches. In addition to water and materials savings, the new facility has led to changes in working practices that provide a better working environment. Annual cost savings of about £220,000 will mean that MacDermid enjoys increased profit with a payback on investment of around two years. More details are available in case study NC260: *Cleaner Technology Brings Cost Savings and Environmental Improvements*.

Amphenol Ltd produces specialist electrical connectors. The connectors are electroplated with conductive and non-reactive metals to prevent corrosion and ensure good conductivity. Electroplating traditionally uses considerable amounts of water and can lead to effluent with a high loading of heavy metals. Amphenol was concerned about its level of water use and the difficulty of meeting more stringent compliance requirements for discharges of cadmium. The company decided to use a closed-loop ion exchange system because it reduced water use, eliminated effluent and recovered metals.

The system that Amphenol installed was very successful, reducing water use by 89 per cent and leading to cost savings of £108,000 in the first year. Despite the high initial outlay the system paid for itself in 18 months. More details of what the company did are available in case study GC24: *Effluent Costs Eliminated by Water Treatment*.

Varian Medial System (UK) Ltd produces radiotherapy simulators, a sophisticated electronic product. The company took a holistic approach to cleaner technology by redesigning its product with the environment in mind. This entailed thinking about the whole life cycle of the product, including dismantling after its useful life. It also entailed working closely with key suppliers to help address some of the important issues they identified.

The approach used by Varian gave significant benefits to its manufacturing process, including considerable savings on materials, components and manufacturing time. The cost of parts alone has been reduced by over £160,000 per year. The design of the machine also makes it easier to recover and reuse parts at the end of its life. In addition to cost and environmental savings, the company has benefited from improved competitive advantage and will be well placed to address forthcoming legislation on end-of-life recovery. In addition to a short case study on the project (NC201) a more detailed report is available (NR201).

Dr Martin Gibson works for AEA Technology Environment as manager of the government's Environmental Technology Best Practice Programme (ETBPP). He has considerable experience of helping companies to reduce waste at source. He has also been active in promoting the business benefits of improved environmental performance and helped to formulate many of the workshops given by the ETBPP team. He has recently published a book entitled *Environmental Sustainability*.

The government's **Environmental Technology Best Practice Programme (ETBPP)** encourages UK businesses to adopt cost-effective measures that reduce waste at source. It offers a range of free, practical advice to help companies do this by, for example, promoting and publishing case studies of successful projects that companies have undertaken. It also produces guides to help other companies make the same improvements. It reaches companies directly and by helping other business support organisations to promote environmental improvements. The Programme has been running since 1994 and has recently been extended until 2007.

The ETBPP has produced a guide entitled *Choosing Cost-effective Pollution Control* (reference GG109) to help companies take a cost-effective approach to adopting new technology. The Programme will be releasing more guidance on how to undertake cleaner design shortly. Information on these, or other publications from the Programme, can be obtained through the Environment and Energy Helpline (0800 585794). The helpline provides free advice on waste minimisation, cleaner technology, or any other environmental issue for companies in the UK. Most publications and the answers to many frequently asked questions can be found on the Programme's website (www.etbpp.gov.uk).

35

The Case for Giving Away A Good Thing

Dick Crosbie,
NIKE

Some things come easy. An idea that springs full blown into your head. A sudden inspiration from something spoken that gives you a vital missing link. Sometimes it's seeing something somewhere else that just rings a bell. It's called being in the right place at the right time.

Unfortunately, most things don't come that way. They come through long, careful, plodding research. Hours of testing and retesting. Results measured in microsteps that lead ever so slowly to a process or product that works, and after all that work you discover that what was so hard to come by was patently obvious. 'Green things' come this way. Green technologies, green processes, green products. None of them comes easily, so why give them away?

There are lots of reasons not to. People will argue that they give you a competitive advantage. Corporations will argue that they paid for the work so it should belong to them. Accountants see savings that could result in a significant rise in profits. For some people it's because they feel they did all the work.

Actually there are lots of reasons for sharing this kind of technology, not all of them necessarily altruistic, although there are of course the altruistic reasons:

- preventing hundreds or even hundreds of thousands of people, plants or animals from possible exposure to toxic substances;
- eliminating or reducing the destruction of irreplaceable limited resources such as water, ozone, oil or land;
- making the world a safer place for our children;
- putting new life into a dying industry;

- helping keep things in the earth's crust and out of the biosphere;
- or just because it makes you feel worthwhile.

But then there are the other reasons. The traditional reasons that your chief executive officer might like. These might include:

- becoming recognised as an industry leader;
- attracting more people to buy your product because you are recognised as an industry leader;
- being able to attract more qualified people to work for you;
- by giving away your technology the whole industry benefits from less attention from governmental regulatory agencies.

Even if one wants to give away something good like a green technology there are always people who are going to make it difficult. There is a maxim that goes: 'All middle management abhors change', and how true it is. There will be those who just don't believe what you are showing them. Others will feel that their grandfather did it the old way, that was good enough for him, so why change now? Still others will feel they can improve upon your work and so set out to prove that you aren't that brilliant – they end up never getting around to implementing the technology. Finally there are those who set out to make the change, but at the first sign of a problem throw up their hands and go back to the old ways. Occasionally there is a manager who embraces change and watching him or her make that change and become a convert is a wonderful thing to behold. In some cases 'the student becomes the master'. However, no technology will ever be successfully transferred without the backing and full support of senior management of the giver or the receiver. Senior management is the key. Convince them and all things are possible.

NIKE Inc has been an industry leader in developing new technologies that we have given away to our competitors. In the early 1990s it was becoming evident that governments were becoming more and more regulatory in how footwear was produced and where it was imported from, by 1992 making the environment and worker health and safety a major corporate goal. Between that time and the present NIKE, in conjunction with its subcontract factories and their suppliers, developed many new technologies – waterbased adhesives and primers for footwear, waterbased mould releases, detergents for outsole degreasing, mould cleaning and as flushing agents. Between 1995 and the end of 1999 NIKE reduced its per-pair usage of organic solvents from 340 grams/pair to slightly more than 50 grams/pair; an 83 per cent reduction. At the same time it developed recycling programmes for adhesive and primer containers. It has reduced scrap levels of EVA and rubber through internal recycling programmes and its Reuse-a-Shoe Program.

In November 1998 NIKE held an open forum in Bangkok for all its competitors on its advances in the area of green technology. All the above technologies were made available to its competitors along with detailed explanations and a complete question and answer period. The forum also included a tour of a NIKE factory that was using a great many of the technologies.

When all is said and done there are many reasons for giving away a good thing. The industry benefits. The workers benefit. The environment benefits. Your company benefits, and hopefully you benefit from the work you've done. There are no losers, only winners. My advice to you is this: if a green technology is being developed try to convince your manager to see the benefits of sharing it; if it's being given away and it's free, by all means try to be the first in line.

Dick Crosbie is the Director of Footwear Manufacturing Chemical Engineering Operations. He has been in the chemical engineering area of the footwear industry for 30 years and has been employed by NIKE for 19½ years – 14 of which were spent in subcontract manufacturing operations in Asia. He has worked in Korea, Taiwan, Malaysia, China and Indonesia. He returned to NIKE's world headquarters, in Beaverton, Oregon, in 1998.

NIKE Inc
One Bowerman Drive
Beaverton
OR. 97005
USA

tel: 1-503-671-6453
fax: 1-503-671-6003
websites: www.Nike.com
*   www.Nikebiz.com*

NIKE is the number one sports and fitness company in the world, with sales exceeding US $8 billion. NIKE sells footwear, apparel and equipment for the discerning authentic athlete. It has subcontract manufacturing sites in Asia, the US, Europe, Cental and South America and Africa.

Section X

Planning Challenges, Contaminated Land

36

Managing Corporate Real Estate: Learn to Stop Worrying and Add to the Bottom Line Instead

Richard Spray,
Blue Circle Industries

Any business – large or small – requires property to function, even e-commerce needs a room for the computer. But how much space, in what form, and where? A primary role of the corporate real estate professional is to assist operational colleagues to achieve the best match between what a business needs in order to function and grow and what it actually occupies.

This sounds both obvious and straightforward, but as one of the essential means of production real estate seldom receives the attention it deserves – at least until something goes wrong. The pressure for profit is ever increasing and environmental constraints grow daily. So allowing property to stand idle or underutilised, particularly where historic activities may have resulted in contamination, is not only bad business – it could very easily put board members in the dock. So who has responsibility for your surplus sites?

Around the turn of the 20th century many industries grew at an exceptional rate through acquisitions and, in the wake of that growth, the value of many assets was not fully realised. In addition they were not subject to the strict environmental, health and safety and planning regulations that prevail today. The emphasis on exceptionally high environmental standards has brought the role of the real estate professional into sharp focus. Expert advice is now essential in evaluating real estate implications of business acquisitions and disposals. Business plans must cover the achievement of minimum standards on property ownership and occupation to ensure best practice throughout the life of an operational unit.

The first and most essential step in the effective management of corporate real estate is to know and understand the extent of your property portfolio, in other words – audit it. Accurate and well-maintained records are critical. The audit should cover everything from legal issues (title, leasehold obligations, etc) to the condition and use of the land and buildings.

Blue Circle Industries achieves this first objective by means of a computer-based property management system. These are available 'off the shelf' and need not be expensive. In Blue Circle's case the property management system is linked directly to a geographic information system that can produce accurate plans on demand. By linking the two together and making them available over the wide area network everyone in the company can have read-only access to the database. The availability of such a database will also permit accurate and realistic key performance indicators to be set and measured.

The next step is easy – understand what your operational colleagues need. This will be different to what they *say* they need and the wisdom of Solomon, otherwise known as the chairman's casting vote, may be needed here. The real estate professional needs to be realistic, however, and be ready to acknowledge that there may be overriding operational reasons for making suboptimal property decisions. In such cases your role will be to ensure that those with the ultimate responsibility for such a decision have all the facts before them – informed decision making is the key!

Once surplus land or property has been identified it becomes the primary task of the corporate real estate executive to manage the fabric of what is left, thereby allowing operational staff to concentrate on operational matters. Secondly, ensure that surplus properties exit the company in the most profitable manner possible consistent with good practice. Increasingly, to exit property profitably means carrying out decontamination or demolition work. So understanding the impact of current and past uses of the site is vitally important. Current uses are managed by ensuring that all major operational sites meet standards applicable to ISO 14001. Past uses are identified by reference to available data such as Envirocheck and by trawling through old records and interviewing employees with past knowledge of the site. Maximising value will also often entail using the town planning system to secure a higher value land use.

The current emphasis on the reuse of brownfield sites, particularly those in urban areas, for residential or retail use means that significant increases in value can be achieved for those willing to invest the time and money in the first instance. Working on this principle Group Property has reduced Blue Circle's UK property portfolio by over 2,500 hectares (10, 000 acres) in the last 10 years and produced a range of valuable redevelopment opportunities.

Blue Circle's most famous example of brownfield redevelopment is of course the 1.7m sq ft Bluewater Park shopping centre built in a worked-out chalk quarry in Kent. This is the largest but by no means the only example of where the company has turned a surplus asset into a bottom-line gain. There are many others schemes, such as the 5,000-home community at Chafford Hundred, and very large commercial and mixed use projects such as Crossways Business Park at Dartford and the 500 homes plus employment space and golf course that make up the Rhoose Point scheme in south Wales. Each one of these schemes provides challenges in achieving marketable planning consents with difficult restoration schemes in an increasingly sustainability-driven environment.

Operational portfolios should not be ignored. Regular maintenance and repair will not only improve their operational efficiency but will increase their capital value. Everyday work pressures on operational staff may mean repairs only get done when they can no longer be ignored – yet few managers would tolerate a leaking roof, blocked drain or rotten window frame in their own home. Why should business premises be any different? Intensive property management is the key to maximising returns to the bottom line and in doing so ensure compliance with the environmental legislation.

Group Property is now helping management throughout the Blue Circle international operation to realise the full potential of landholdings and to ensure minimum standards are set in decommissioning and reclamation activities. There are common factors to real estate management worldwide and we are able to identify and direct local management to ensure that decommissioning, reclamation and disposal strategies exceed environmental, planning and local community expectations.

One final golden rule – never buy or lease property without establishing your exit strategy. This will change over time, but having one in the first place will keep you ahead of the game when the inevitable day comes to move on!

In summary, property must be made to work just as hard as any other aspect of a business. Ignore it and it will cost you money, manage it properly and there is nothing to fear, and there is the strong likelihood of significant gains to the bottom line.

Richard A Spray is Head of Group Property, Blue Circle Industries plc. He was Royal Institution of Chartered Surveyors qualified in 1971 and started his professional career in the same year with Slough Estates, where he received a good grounding in estates management. In 1974 he joined ARC Ltd, the aggregates company, initially as management surveyor then as South-east Regional Estates Manager and finally as one of the founder members of ARC Properties, bringing forward development opportunities on surplus aggregate sites.

He joined Blue Circle in 1991, and his duties include managing the operational estate and once again maximising the value of the surplus estate. Richard says: 'From the above you can see I have spent nearly all of my professional life in the sexy side of the property industry – grit, rock and cement!'

Blue Circle Industries is now a focused heavy building materials company operating globally, having sold its bathrooms business. It is currently disposing of its pan-European heating products division.

Blue Circe's land ownership is around 25,000 acres in the UK, the same again in North America, approximately 7,500 acres in Chile, 4–5,000 acres in Greece and a portfolio size in Africa, Malaysia and the Philippines which is unknown at this time. Note that the company deals in acres and hectares, not sq ft or metres – it is a land-hungry business.

The Group is vertically integrated from gravel, ready mixed concrete, concrete products through to cement. It is UK based, and about 90th in the FTSE100 with a capitalisation of some £3.7 billion. It is probably fourth or fifth largest cement producer in the world, with approximately 20,000 employees.

Blue Circle Industries plc
84 Eccleston Square
London SW1V 1PX

website: www.bluecircle.co.uk

37

Contaminated Land and its Liabilities

Tony Allen,
Donne Mileham Planning

Introduction

The Part 2A Environmental Protection Act 1990, which enables local authorities (LAs) or the Environment Agency to require the remediation of contaminated land, came into force on 1 April 2000. Previous legislation, Section 143 of the Environmental Protection Act 1990, was rejected because of unpopularity with the development industry and others; the present code was introduced as part of the Environment Act 1995, after which there followed a five-year process of publishing and revising draft guidance on the detailed operation of the scheme.

Estimates of the total area of contaminated land in the UK vary from 16,000 hectares to 200,000 hectares; whatever the true figure remediation has become more urgent as development plans concentrate increasingly on previously developed land, and tightening environmental controls make it essential that problems caused by contamination are quickly and effectively dealt with.

It is important to note that the new legislation will only deal with land where contamination is causing 'harm' (as defined) or pollution of controlled waters, or is likely to do so. If contamination can be shown to be contained within a site this legislation will not apply – a distinction to be borne in mind during property transactions.

The new legislation

The legislation gives powers to local authorities, or the Environment Agency in the case of special sites. These include those likely to be affected by serious contamination because of previous occupiers, eg military bases, where the ground is known to be contaminated by difficult substances, or where the site overlies certain identified geological formations.

Local authorities must publish, within 15 months from 1 April 2000, a strategy for purveying and securing remediation of contaminated sites in their area, to ensure that the problem is approached in a logical way, taking the worst sites first.

For land to be identified as 'contaminated' the LA must demonstrate that, by reason of the presence of substances in, on or under that land, significant harm is being caused or there is a significant possibility of such harm being caused, or pollution of controlled waters is being or is likely to be caused. This will require the LA to identify a **contaminant** in the site, a **receptor** outside the site and a **pathway** by which the contaminant is reaching or is likely to reach the receptor. These three factors will together amount to a 'pollutant linkage'. The LA will then need to demonstrate that the risk of harm, etc, is significant to identify a 'significant pollutant linkage'. It is then in a position to consider whether remediation is required and, if so, what form it should take and who should carry it out.

The guidance requires an LA, having identified contaminated land, to give those likely to be required to carry out remediation a three-month period within which to discuss with it, and if necessary among themselves, the form remediation should take and who should carry it out, and the allocation of remediation costs. Only at the end of this period will the LA be entitled to serve a 'Remediation Notice' by which it will formally specify the work to be done, the party to do it, etc.

Who may be liable

The persons who are or may be responsible for remediation are entitled **'appropriate persons'** in two categories: Class A, those persons who have caused or knowingly permitted the contamination, which may include those who, being aware of its existence, had the opportunity to carry out remediation but had failed to do so; or Class B, those who are owners or occupiers of the land, not having been responsible for the presence of the contamination.

The rules provide that Class B appropriate persons can only be responsible for remediation if no Class A appropriate persons can be found. The rules for identification of appropriate persons, and the sharing of liability between them by reference to causation, the nature of contaminants in the land, etc, are complicated and are likely to be tested extensively as the legislation comes on stream and LAs begin serving Remediation Notices.

There are provisions to exclude from liability persons involved with land solely by virtue of, for instance, the funding of its purchase, insurance, the provision of legal services, landlords (where tenants have caused contamination), the provision of engineering, scientific or technical services and various other categories, subject to the proviso that the party in question has not been directly involved in management of the site or the causing of the contamination.

Controlled waters

In relation to 'controlled waters' the rules are different. There is no need to prove *significant* contamination – any escape to controlled waters is sufficient for the LA to take action. There is also an overlap with the provision for Works Notices under Section 161A of the Water Resources Act 1991, but recent guidance is to the effect that LAs should normally use their powers under the Environmental Protection Act.

Practical effects

In practical terms it would appear that land identified as contaminated may be 'remediated' by relatively simple measures sufficient only to break the contaminant–pathway–receptor linkage, rather than complete clean up of the site. The legislation and its guidance emphasises that remediation will only be required to a 'fit for use' standard. The legislation also provides that it should only be used as a last resort, and that if, for instance, land is to be redeveloped, remediation should be secured through the planning system (by planning condition, Section 106 Agreement or other suitable mechanism) rather than by the use of these powers. This seems likely to result in a situation where different local authority officers will be responsible for requiring and supervising remediation depending on the use to which the site is to be put. It is also clear that different LAs may adopt different approaches to the implementation of the legislation.

The legislation makes provision for appeals against Remediation Notices, for which the procedure would be similar to planning appeals, the carrying out of remediation by local authorities or the Environment Agency where the appropriate person defaults, and the recovery of costs from the appropriate person and for the keeping of registers of contaminated land. These will be particularly significant because once a site is entered on a register there is no provision for it to be removed, even if remediation is carried out to the satisfaction of the LA. Registers will generally be open to public scrutiny, although there are provisions for exclusion in the case of national security or commercial confidentiality.

Great care will need to be taken in property transactions to ensure that contaminated sites (whether within or outside the new legislation) are identified and responsibility for remediation costs allocated between the parties.

'Sold with information'

The guidance provides that, in transactions since 1 January 1990 involving **large commercial organisations** or **public bodies** (which terms are not defined) liability for contamination in land will be deemed to have passed to the purchaser if that purchaser was given a reasonable opportunity to acquire information about the contaminated state of the land before being committed to the purchase of it. Present opinion is that a sentence in a letter between solicitors offering the purchaser the opportunity to inspect the site, whether or not this was taken up, would be sufficient. There may therefore be a number of cases where purchasers have inadvertently become liable for remediation of contaminated sites.

Insurance

The insurance market has responded to the risks arising from contamination and is offering a number of products involving cover against remediation and other costs.

Houseowners

While the new legislation may appear to be mainly of concern to the owners or occupiers of industrial and other similar sites, it does also pose risks for houseowners, where properties have been built on land where remediation has not been carried out adequately or at all, and the developer or other previous landowners have 'disappeared'. The National House-Building Council (NHBC) has added cover for purchasers of new properties in recent transactions, but those who have acquired their properties earlier could find themselves caught as 'appropriate persons'.

Changes in standards

One other concern is that, over time, remediation standards may change, so that a site which is cleaned up to an agreed standard today may require further work in the future if acceptable limits for the contaminants in question are reduced or new substances are identified as contaminants.

Comments

As will probably be apparent, the new legislation is very complex and in places badly drafted; some local authorities are struggling to cope with the requirements it places on them. In many property transactions questions relating to land contamination will or should be high on the agendas of the parties, but it may take some time for a clear picture to evolve as to how the legislation will actually operate and impact on landowners or those who have been responsible for past contamination.

Tony Allen is an environmental law specialist with Donne Mileham solicitors. He leads the firm's substantial in-house planning and environmental law group, which includes both chartered planners and solicitors. He is the author of a recent book on the new contaminated land regime (copies of which are available from his office). He and his team can advise on most environmental law issues.

DMH Planning
100 Queens Road
Brighton
East Sussex BN1 3YB

tel: 01273 744450
fax: 01273 744455
e-mail: tony.allen@dmh.co.uk

Donne Mileham is one of the leading legal firms in the South-east, providing a diverse range of legal, planning and financial services to commercial and private clients.

   The firm's 28 partners and 220 staff are organised into specialist teams with in-depth knowledge and relevant experience in their various areas. Each team works closely with the teams in related fields to deliver a broad but integrated service tailored to the requirements of clients. The firm's main areas of expertise are employment law, litigation, corporate finance, commercial property and property development, planning and environment, financial services, business and personal taxation, intellectual property/IT, company commercial, technology, media and telecommunications and the not-for-profit sector.

   The *Chambers & Partners Guide to the Legal Profession* currently names 11 of the firm's lawyers as leaders in their respective fields.

38

Planning Restrictions and their Impacts on Businesses

Pat Thomas,
S J Berwin

What to look out for when taking new premises or looking at land to develop new premises

Man has not had an inalienable right to do on his land what he wants for centuries, but the last century and increasingly this century have imposed ever more restrictions. It is often difficult to remember that these restrictions are meant to protect the public interest. They are good for you.

The development plan – its importance

Any proposal for new development is likely to require planning permission, and how the local council, which in most circumstances will be the local planning authority, determines whether to grant planning permission is regulated by the terms of the development plan for the area where the development site is located. The government has emphasised the importance of ensuring that there is nationwide coverage of up-to-date development plans, and has ensured that sufficient resources are available through the DETR to make this happen.

The need to contribute to the development plan review

This means that where a business has plans to expand it should monitor the reviews of the development plan to make sure as far as practicable that it reflects its requirements. Land adjoining an industrial complex may be re-designated for housing without objection from the adjacent business, so that when that business seeks to expand or intensify its operations by installing new plant or buildings such development is deemed to be unacceptable next to the new family housing recently constructed in accordance with the development plan.

Keeping in touch

This monitoring is achieved through staying in touch with the local planning authority so that the officers in the Planning Department responsible for preparing reviews of the development plan know the requirements and aspirations of the local businesses which will generally be the source of the main employment in the area. Information on policy changes and the co-ordination of objections and representations is regularly carried out by the CBI regional offices through liaison with Centre Point.

Existing buildings will generally have been constructed under a specific planning permission. This may not apply to buildings over 50 years old which were constructed before the current town and country planning regime came into force, but separate statutory provisions will then apply to establish the status of such buildings. Where there is a specific planning permission it will normally have been granted conditionally. These conditions will have been imposed to protect the general good and effectively amount to restrictions on the development. Conditions can limit the amount of floorspace to be created by the development and the use of that floorspace. For example, it is usual to find conditions on major retail developments restricting the amount of floorspace to be devoted to comparison goods and/or convenience goods in order to protect the viability of existing retail development. In town centres conditions may be imposed restricting the hours of operation. The industrial concern that has a new residential development on its boundary may find that any planning permission granted for expansion has conditions restricting open storage, hours for delivery, use of reversing klaxons, etc, all directed to protecting the amenities of the nearby residents, without much regard to the viability of the business where such restrictions are imposed.

An informed local planning authority can be a friend

Before imposing such restrictions, or where a business seeks to have such restrictions on pre-existing planning permissions relaxed or removed, the local planning authority, and the Secretary of State for the Environment, Transport and the Regions on an appeal, should weigh the benefits of having a thriving local economy with growing businesses seeking to increase their employment potential against the subjective insistence by local residents upon noise controls, traffic bans, etc. In order to be able to carry out this assessment the local planning authority needs the comprehensive information that only the local businesses can supply. Such is the workload in some local authorities that it is easier to say 'No' than to spend time chasing up the information from the commercial sector. The local residents, however, will always find time to make their objections known.

Third parties and the Human Rights Act

Their hand has been strengthened by the Human Rights Act 1998 which recognises in Article 8 the right to respect for private and family life, and in Article 1 of the First Protocol, the right of a person to the peaceful enjoyment of his or her possessions. These rights may only be infringed in certain circumstances, which in relation to development can include the economic well-being of the country and the general interest. The acknowledgement of these rights also means that planning officers when recommending their committees to grant planning permission will be anxious not to infringe such rights or to mitigate any infringement, by imposing restrictions on any planning permission or through obligations in so-called voluntary development agreements. Again, the only recourse for businesses is to remain vigilant and to ensure that the economic development or planning officers in their local authorities are aware of the economic rationale for the development proposals, and the impact upon the commercial operations of too restrictive an approach, provided that public health or the sustainable environment is not put at risk.

Pat Thomas has 25 years' experience of planning and related matters both in the private and public sector. After spells in the public sector and a leading City practice, she moved to S J Berwin. As Head of the Planning and Environment Group she leads a team of dedicated specialists advising on all aspects of planning and environmental law including highways, land drainage and flood prevention, historic buildings and nature conservation. 'Now widely regarded as the City's leading planning lawyer' (*Planning*, May 2000).

S J Berwin & Co was established in 1982 with the fundamental objective of providing outstanding legal advice in a dynamic environment, and the practice has built a reputation that is the envy of many longer-established firms. With offices in London, Munich, Brussels, Frankfurt and Madrid, the firm is unhampered by bureaucratic attitudes, operating in the type of open, accessible and fast-moving atmosphere that promotes progressive thinking and a creative approach to meeting clients' needs.

S J Berwin & Co
222 Grays Inn Road
London WC1X 8XF

tel: 020 7533 2222
fax: 020 7533 2000
e-mail: info@sjberwin.com
website: www.sjberwin.com

Measuring and Reporting Environmental Performance and Stakeholder Dialogue

39

Why Failure to Report is a Risky Business

John Elkington, SustainAbility

The 1990s saw the rapid evolution both of what some called the 'CNN World' and of voluntary corporate environmental reporting. Business people, as politicians before them, have found themselves having to operate in conditions of unparalleled transparency. Some companies, concluding that they might as well put their own version of the facts into the public domain, decided to report publicly their challenges, dilemmas and progress. Early pioneers included Norway's Norsk Hydro and the US company Monsanto, of which more in a moment.

In retrospect, this new corporate appetite for disclosure was an extraordinary step forward. For decades business had fiercely resisted demands for greater corporate transparency in such areas as business ethics, environmental and social performance, and other societal priorities. More recently the market's growing demand for social responsibility and risk-related information has given the corporate disclosure trend a real boost – and will continue to do so. But this is not the only energy now working to flip the predominant paradigm from 'closed' to 'open'.

A succession of corporate controversies has rocked major companies such as Shell (Brent Spar and Nigeria), NIKE (social conditions of labour in Asia) and Monsanto (genetically modified foods). More recently still, the World Trade Organization was rocked on its heels by protests in Seattle, with growing pressure on it to pay more attention to social and environmental issues. Such evidence suggests that people in developed and developing countries alike are increasingly concerned about a future shaped by globalisation, capitalism and giant corporations.

As national governments seem to lose whatever is left of their power to hold key parts of the private sector accountable, the question arises as to who should design the necessary global checks and balances – and how. Democratic, sustainable global governance systems are urgently needed. Business people who ignore these signals risk seeing the global economy collapse back into nationalistic, protectionist regional enclaves.

That's part of the big picture on corporate accountability. But, at the national level, we are also seeing corporate governance inquiries producing interesting recommendations in this area. The recent UK Turnbull Report on corporate governance suggests that boards and directors will need to understand risk management in a much broader context. This impetus should help those trying to encourage boards to think about sustainable development's 'triple bottom line', which requires companies to focus not only on the economic value they add but also on the environmental and social value they add – or destroy.

Already the centre of gravity of the sustainable business debate is shifting from public relations to competitive advantage and corporate governance – and, in the process, from factory fence to the board room. The new definition of corporate risk will oblige boards of directors to address such issues as health, environment, human rights and business ethics. On the international front, they will need to apply the same standards for joint ventures and other partnerships.

Meanwhile too many businesses – among them many internationally known brand name companies – are choosing to remain silent on their environmental records and impacts (see *The Non-Reporting Report*). For example, less than 10 per cent of the companies that have signed the International Chamber of Commerce's Business Charter of Sustainable Development have actually reported, even though they are encouraged to do so by the Charter. As a result, governments will be forced to get tough with non-reporters. Those that have recently threatened to 'get tough' with business include the Netherlands, Sweden and the UK. UK Environment Minister Michael Meacher has repeatedly warned companies that if they do not respond to the carrot of voluntary initiatives the government will have to resort to the stick of new regulations.

So why, when over 600 international companies have produced at least one environmental report, and some 2,000 EU industrial sites have produced so-called EMAS reports, do so many companies remain silent? One answer, we have found, is that many non-reporters say they see little benefit in reporting, although reporting companies have often been surprised by both the internal and external benefits. For others the lack of a common reporting framework makes reporting a more onerous – and ultimately less meaningful – exercise.

SustainAbility and the United Nations Environment Programme (UNEP) are now tackling this barrier through a new series of sector-based disclosure initiatives, designed to catalyse a core set of common indicators for particular industry sectors. Early projects have covered the oil and life sciences sectors.

At the same time, the new Global Reporting Initiative reporting guidelines offer a useful framework for companies wanting to produce state-of-the-art reports (see their website at: www.globalreporting.org).

But an even more powerful pressure is now evolving as major companies start to challenge their suppliers in such areas as health, safety and environment. When a company such as Ford certifies all its sites, worldwide, to the ISO 14001 environmental management standard, as it recently did, we can be sure that the company will also want to shine the spotlight up and down its supply chain. It knows that activists are already doing the same – and that in these days of the Internet bad news can break far faster than even the best-organised companies can sensibly respond.

Indeed the main argument for reporting is internal, not external, to companies. Although they may start off wanting to convince their critics that they have the main issues under control, many soon find that the real benefits flow from a better understanding of the existing and emerging risks, the associated review of internal management systems and processes, and the preparedness to answer key client questions in a timely and convincing fashion. In the 'CNN World' companies that fail to report in the context of new definitions of corporate accountability may – unwittingly – be playing a commercial version of Russian roulette.

John Elkington is Chairman of SustainAbility and co-author of over 10 reports on environmental, social and sustainability reporting, including *The Non-Reporting Report* (SustainAbility and UNEP, 1998). He is author of a number of business books, including *Cannibals with Forks: The Triple Bottom Line of 21st Century Business* (Capstone, 1997) and *The Chrysalis Economy: How Citizen CEOs and Corporations Can Fuse Values and Value Creation* (Capstone, 2001).

SustainAbility works mainly with the private sector, its clients including Ford, Holderbank, ING, the International Finance Corporation, Shell International and the United Nations Environment Programme. It publishes reports on sustainable development issues.

SustainAbility Ltd
11–13 Knightsbridge
London SW1X 7LY
tel: 020 7245 1116
fax: 020 7245 111

e-mail: info@sustainability.co.uk
website: www.sustainability.co.uk

40

An Approach to Environmental Reporting: Anglian Water

Ken Smith,
Anglian Water

The foundation of the environmental report

Anglian Water, a FTSE100 company, is a major provider of water and environmental services to 5 million customers in the east of England and to over 4 million customers around the world. Water is taken from the environment, treated and supplied to customers. Their wastewater is then collected, treated and returned back to the environment. Playing an integral part in the water cycle means that environmental protection is at the heart of the business.

With privatisation of the industry in 1989 Anglian Water made this close link to the environment explicit with the publication of a corporate environmental policy in 1990. A statement of intent described in a corporate environmental policy is a good place to begin as a foundation for a public environmental report. For Anglian Water, the policy is the framework around which the annual environmental reports are built.

The first line of our environmental policy is to 'meet fully our legal obligations to the environment' and so fundamental to the report is our progress against statutory duties and environmental legislation. Along with this is a description of the managerial processes and procedures that are required to ensure compliance. Over the years interest in a wider range of environmental issues that go beyond regulatory duties has increased, both within the company and also externally with our stakeholders. The annual environmental reports have therefore grown to reflect these needs. But it is essential when reporting on statutory duties and wider environmental issues that there are measures and targets to allow progress to be monitored. Looking again to our environmental policy, we state our aim to 'measure our performance against specified target'.

The audience

A key issue to consider before writing an environmental report is: 'Who is the audience for the report?' And this is difficult, as it is natural to want to write a report for all stakeholders – the style of writing, content and 'story' that is in a report aimed at a customer audience is very different from that for opinion formers and regulators, for example.

Anglian Water's annual Environmental Activity Reports are specifically aimed at opinion formers, regulators, environmental groups, investors and UK industry. To meet the needs of this audience we also decided that the environmental report would provide a rounded insight into all our investment and activities that contribute towards sustainable development. Therefore the three main facets of sustainable development – financial, environmental and social – are blended together to give an overview, have separate sections to allow readers to access specific information of interest, eg the performance, and a compliance section for investors and regulators.

However, there is no point in substantial duplication in different company annual reports - such as environmental reports, and statutory annual reports and accounts. Clear 'signposting' has therefore been provided in all reports to allow cross reference to where greater detail can found – including the Internet. This also applies to our summary Environmental Review for customers and schools, where there is reference to the main Environmental Activity Report for the detailed information.

The aim and use of the environmental report

Defining the audience goes a long way to defining what is to be gained from publishing an environmental report. For Anglian Water the report is aimed at clearly defining the company as a responsible organisation – in the way that money is invested in meeting the needs and improving the quality of life of customers while protecting the environment. The main environmental report therefore is used to differentiate the company:

- with opinion formers – government, pressure groups, regulators, MPs;
- with investors showing effective deployment of finances to exceed customers' expectations;

- in the bidding process for overseas contracts;
- in recruitment and induction of new employees and also as a training/
  reference tool in environmental awareness sessions;
- with further education establishments teaching environmental courses.

The summary Environmental Review serves the same purposes with customers and schools.

An evolving report

Each year when the Anglian Water report is published it is necessary to start
thinking about the next report and how it can evolve and improve to ensure
readers' expectations are met. One way is to read environmental reports from
other companies, but probably the most important is gain feedback directly
from the readers. In previous years we inserted a reply-paid response card
but this was quite expensive with a low response rate of around 5 per cent –
although this information was still useful. Now we use several methods to
gain feedback:

- My personal e-mail address is given in the report so readers can respond to
  a named individual rather just to the organisation, and I respond to all e-
  mails.
- A simple response form on the Internet with the report, and all these
  responses are answered.
- Direct one-to-one consultation with the audience we are targeting. For two
  years now we have consulted with over 30 opinion formers from a wide
  range of interests and asked about content, design, etc, which has provided
  valuable information that has allowed the report to be improved.

These methods have led to a response rate of over 10 per cent at very low cost
and also I believe have been one of the reasons that Anglian Water was the
overall winner in the 1998 Association of Chartered Certified Accountants
Environmental Reporting Awards – for which we had been shortlisted for
several years. Winning the award has in itself generated significant demand
from around the world for the report, with several hundred copies being
requested and also several hundred responses from individuals downloading
the report from the Internet.

Conclusion

The approach to environmental reporting that we have taken at Anglian Water
is:

- base the report around the corporate environmental policy with clear objectives and targets
- define the aims and audience for the report
- gain feedback from readers so that their expectations are met
- keep it simple!

Ken Smith is the senior scientist for environment and conservation policy. He has worked with Anglian Water for two and a half years but has been involved in environmental management for over five years. His background is in environmental science, ecotoxicology and environmental management and auditing.

Anglian Water
Environmental Affairs Team
Endurance House
Histon
Cambridge CB4 4ZR

tel: 01223 547559
fax: 01223 547566
e-mail: ksmith4@anglianwater.co.uk

Anglian Water is the largest water and wastewater company in England and Wales, serving over 5 million customers from the Humber in the North to the Thames in the South and from the East coast to Daventry in Northamptonshire. Water and wastewater services are provided to over 4 million customers outside the UK in countries such as New Zealand, China, Chile and the Czech Republic. The Group turnover in 1998/99 was £829 million.

website: www.anglianwater.co.uk

41

Rewarding and Encouraging Reporting

Roger Adams,
Association of Chartered
Certified Accountants

Environmental reporting: enriching corporate governance

The meteoric rise in recent years in the number of companies producing separate environmental reports and environmental information in their annual reports and accounts reflects the growing concern many companies now have for the environmental degradation they inflict. It is also another convincing argument for non-reporters to acknowledge their impact and take responsibility for it.

Many companies have recruited managers to oversee corporate environmental affairs, have delegated environmental responsibilities to board members, and created environmental committees from a cross section of its departments (with some having the additional benefit of expertise from external advisers). Executing good environmental practices within corporations and reporting on them externally is now an integral part of effective and successful corporate governance.

Two recent publications, *Company Law Reform* (DTI), and *Internal Control: Guidance for Directors on the Combined Code* (the Turnbull Report), should convince the most cynical of companies that environmental risk measurement and reporting issues are being seriously addressed by bodies that historically have always had a financial focus.

Business in the 21st century

Increasingly, companies are realising they need to change the way they 'do business', by becoming more open and accountable for their activities, in order

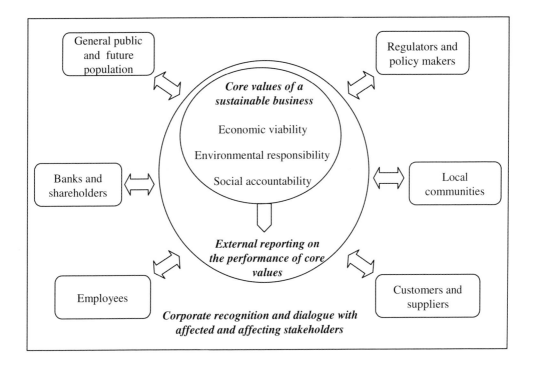

Figure 41.1 *Corporate governance for the 21st century*

to meet the growing demands made of them from stakeholders and government. Companies need to identify and communicate with their stakeholders, to consider and act on their needs, and to involve them fully in corporate business. Companies must also begin to fully integrate economic viability with environmental responsibility and social accountability (core values), to make their operations more sustainable. To complete the picture, external reporting is necessary to communicate the company's performance on its core values to all stakeholders.

The awards

The Environmental Reporting Awards, founded in 1991 by the Association of Chartered Certified Accountants, has significantly improved the standard of environmental reporting in the UK. Each year the panel of judges for the Awards scheme produces a report highlighting current best practice in environmental reporting. The objective of the ERA is to identify and reward innovative attempts to communicate corporate environmental performance, but not to report on good performance itself. The panel of judges, which reflects the range of stakeholder interests, has evolved a set of criteria for reviewing the quality of environmental reports.

Companies that have produced relevant, reliable, complete and verified information in their reports have been rewarded with the Award. The ERAs are now a major national initiative, reflecting the growth in corporate environmental reporting and increased demand from stakeholders for corporate environmental accountability. The Awards scheme has proved influential in the development of corporate environmental reporting around the world, and has been mirrored in many other countries, for example Denmark, Germany and Japan.

Pressures to report

Since environmental reporting in the UK is a purely voluntary activity, it is necessary for companies to perceive some tangible benefits when establishing the business case to report. Some of the benefits most commonly cited include:

- demonstrates coherence of overall management strategy to important external stakeholders
- reduces corporate risk that may reduce financing costs and broaden the range of investors
- effective self-regulation minimises risk of future regulatory intervention
- increases competitive advantage (the 'first mover' effect)
- strengthens stakeholder relations
- public recognition for corporate accountability and responsibility
- may improve access to lists of 'preferred suppliers' of buyers with green procurement policies
- enhances employee morale.

The government wrote to the FTSE100 companies in 1998 requesting that they start reporting on their environmental performance if they had not already begun to do so. This initiative was soon extended by writing to the FTSE mid 250 requesting the same action and there are plans to extend the list to all 7,000 UK companies with more than 250 employees. Mandatory reporting has been hinted at if the DETR is unsatisfied with the outcome.

Innovations in reporting

The lack of generally accepted environmental reporting standards has led companies to develop their own unique methods. As demonstrated in the box, there is a variety of different approaches currently in use. In the UK the most common reporting methods appear to be a mix of compliance and performance-based approaches.

Reporting methodology	Description
Compliance-based reporting	Reporting the level of compliance with external regulations and consent limits is often the core feature of the environmental reports of heavily regulated utility industries such as water and electricity.
Toxic release inventory-based reporting	Many US companies are required by law to publish lists (detailed in physical quantities) of emissions of specific toxic substances. These mandated disclosures often take precedence over voluntary performance-based disclosures.
Eco-balance reporting	Some companies (including many from Germany) construct a formal 'eco-balance' – a detailed account of resource inputs and outputs (in terms of product output and waste/emissions) – from which they then derive performance indicators.
Performance-based reporting	Perhaps the most common form of environmental reporting. Reports are usually structured around the most significant areas of environmental impact. Performance improvement targets are then set and appropriate performance indicators developed and disclosed annually.
Environmental burden reporting	The UK chemicals company, ICI, has developed an externally focused reporting approach that quantifies the company's impact on several environmental quality measures. A potency factor allocated to each emission is multiplied by the quantity emitted per year to provide the environmental burden.
Product-focused reporting	Volvo has produced an 'environmental product declaration' report that evaluates the total environmental effect of one specific Volvo model. Issues covered include operation, recycling, manufacturing and environmental management. In this case, environmental responsibility has extended beyond the factory gate.

The benchmark was progressed during February 1999. In all 11 production companies took part, located in the US, Canada, Mexico, the UK, France, Spain and Portugal. Participating companies received copies of the questionnaire, which comprised 75 questions, looking at a number of EHS issues: organisation and culture, management systems, environment, health, safety, transport, product stewardship and external influence. Each questionnaire was completed by an on-site team of around eight people drawn from all parts of the business and at different employee levels. The team discussed and agreed the scores for each of the questions. Where questions had a UK flavour participants were asked to insert their own corresponding national situation and to respond to the question accordingly.

Following the completion of the questionnaire a day was then agreed for the facilitation visit. This involved the facilitator meeting the team and reviewing its scores. In the UK, CBI Contour facilitators are able to give immediate feedback, which is also supplemented by an individual report showing how the business is performing. For the Allied Domecq benchmark three members of the International H&S Co-ordination Committee were nominated as the facilitators and the completed and facilitated questionnaires were sent back to the CBI for evaluation.

The results of the benchmark comprised a graphical series of performance versus practice charts that covered an overall EHS viewpoint and also individual charts for environment, health and safety, and organisation and culture. A quartiles index plot for each of the main EHS issues covered by the questionnaire was also prepared, together with a written summary covering the position of each participant and potential areas for improvement. Each chart was split into six sections: world class, contenders for world class, promising, could do better, vulnerable and finding your feet. The overall practice versus overall performance chart for all 11 Allied Domecq sites that participated showed 7 as 'contenders', 2 as 'promising' and 2 as 'could do better'.

Results were presented by Jenny Barker to the Allied Domecq International H&S Co-ordination Committee during April 1999. Since then participants have fed back the benchmark findings to their local teams and have used the results to prepare individual targeted action plans to achieve continuous improvement.

With regard to the environment, Jan Buckingham – Director of Alcohol and Social Policy – and Jean François Lavigne – who is the Director of Environmental and Technical Affairs and also one of the facilitators – have responsibility for setting policy and ensuring it is implemented throughout the Allied Domecq group of companies. They both advise the board of Allied Domecq via Richard Turner the nominated main board director with environmental responsibility.

The last 12 months have seen a number of significant changes in Allied Domecq, both structurally and in the approach to managing the impacts of our business in the environment. The work focus has been in implementing a systematic approach to environmental management.

Since its conception the Allied Environment Policy has had as a minimum the requirement for all operations to meet all relevant legislation and regulations. This policy has been extended to include meeting Allied Domecq's own set of minimum standards where legislation does not set adequate environmental safeguards. The main element of the new system has been the selection of ISO 14001 as the production standard for all company operations. ISO 14001 is the only internationally recognised environmental management standard and publicly commits the company to a process of continual improvement.

Allied Domecq's *Environmental Performance Report 2000* explains the timetable for the company to achieve ISO 14001 certification across the group and gives more details of the system and its requirements for data collection, training, reporting and improvements against agreed targets. This commitment to challenge ourselves to make continual improvements and report on progress to the satisfaction of an external verification process is the most significant step made by the company in meeting its environmental responsibilities in the 1990s. Future environmental reports will chart our progress and allow all interested parties to judge our record of performance.

By participating in the CBI Contour programme and extending it beyond the UK to all our international operations we are able to effectively benchmark our performance against an agreed set of criteria. In Contour we have a powerful tool both to assess our performance against top companies and to drive improvements internally by identifying best practice.

The next step will be to undertake the benchmark again during April 2001. The key value of the exercise so far has been the discussion and interaction that has taken place between the companies. The health and safety committee would welcome a truly international questionnaire that would overcome the problems encountered with the questions that were UK specific.

In conclusion, the Contour benchmark exercise has been successful and is enabling Allied Domecq to make an EHS difference in a variety of cultures, with country-best performance moving the company as a whole towards continuous EHS improvement.

Bob Urwin is Group Health and Safety Adviser at Allied Domecq.

Allied Domecq plc comprises the world's second largest international spirits group and a leading global food service business. An international, branded business, new Allied Domecq came into effect in September 1999 following the disposal of the company's UK retail business. The spirits and wine business owns 12 of the top 100 premium spirits brands, including Ballantine's scotch whisky, Beefeater gin, Kahlúa and Sauza tequila. The business is backed by a comprehensive owned or controlled distribution network. Successful and leading brands feature in our quick service restaurants business, which include Dunkin' Donuts and Baskin-Robbins.

Allied Domecq plc
The Pavilions
Bridgwater Road
Bedminster Down
Bristol BS13 8AR

tel: 0117 978 5000
fax: 0117 978 5300
website: www.allieddomecqplc.com

43

Should Business Care? Facilitating Stakeholder Dialogue

Sandra Palmer,
United Utilities

Why should business care about its impact on society? Business only takes action if there is a business case.

Should business care:

- If young people don't have the skills that business needs?
- If the climate is changing and its getting hotter or colder?
- If customers care about the business impact on the environment?
- If customers cannot pay their bills?
- If customers have extra needs?
- If regulators, government and stakeholders demand action?
- If the business affects public health and safety?
- If business is missing the opportunity to improve its reputation or gain competitive edge?

We can't ask these questions and then conclude that business shouldn't care.

Every business has an impact on the society in which it operates. Actively managing that impact is just common sense. It is not an original, nor indeed an entirely creative thought, but it is true. The best programmes in this area to date embrace good management disciplines, passionate personal commitment from people across the business, and a readiness to account openly for their impact.

Accounting for business impact on society is still at an early stage. Companies are required by law to produce annual reports that look at their financial performance. Many also produce regular environmental reports, but increasingly companies are starting to tell their stakeholders how they are performing in the potentially more complex area of their impact on society – and telling the story warts and all.

As a multi-utility United Utilities has a major impact on society and it makes sense to manage that impact. It's like any other management discipline. It also makes sense for United Utilities to report on it – we have an extremely good record and there are increasing expectations among our customers, regulators and government for us to be accountable.

What is this 'impact on society'?

United Utilities is a company operating worldwide. A few figures will give the scale of our impact. In the north west of England (our major base) we have 7 million customers receiving water services and 4.7 million customers receiving electricity services. To deliver those services we have 10,000 employees. In fact we are probably the biggest employer in the North West. We provide essential daily services. We keep the taps running, the toilets flushing, the lights switched on and many of the telephones ringing. We keep people safe and warm. Clearly, because of this, our business has a major impact on the communities in which we operate.

At its simplest it is about ensuring that our water is safe to drink, that our works don't smell and that our electricity network is reliable and isn't an eyesore. But, as you would expect, it is about much more than this. It is about our policy and how we treat our customers, our employees and our suppliers, as well as the environment and the regional economy.

Why do we bother, you might ask. In some cases we have no choice. We are required to do some things by government and by the regulators that give us our licence to operate. But the majority of our actions are voluntary. We are taking them because we want to meet the expectations of our stakeholders that we are a company which has regard for its communities and its environment. We understand what our stakeholders expect of us.

Business has a choice in engaging stakeholder opinion. It can follow a rainbow of options, from totally ignoring stakeholders' opinions to actively engaging them in implementing solutions to business issues. United Utilities is part way along this rainbow in engaging with stakeholders through individual consultation and focus groups and then acting accordingly. The results of our consultation clearly show that our stakeholders recognise the importance of a socially responsible approach. Overall 89 per cent say that a positive record on social responsibilities makes a difference to reputation. In our latest, 1999 consultation, 75 per cent agree that United Utilities meets or exceeds their expectations on social responsibilities (compared to 54 per cent in 1997).

Every business, whatever it produces or whatever services it provides, has an impact on the society in which it operates. What we believe at United Utilities is that actively managing that impact is common sense. It is an essential ingredient in our long-term strategy to be a successful business with a positive reputation.

United Utilities published its first Social Report in 1999. With the Environment Report and the Annual Review this completed the triple bottom-line commitment to holistic reporting on our social, environmental and financial performance. Our first Social Report is both an end and a beginning. It completes a two-year strategic review of our impact on society and opens the issue to broader debate. It reports on the actions we've been taking, how we work in partnership with the community and how we measure our progress. Our two-year strategic review included checking alignment with our business objectives and understanding the views of our stakeholders.

We now have a clear policy that we deliver through our brand companies. There are no simple answers to some issues of social impact. So throughout our report the reader is offered signposts to further information and opportunities to debate in dialogue. This is designed to push some of the debates on to the business agenda. Issues such as 'should we have statutory legal standards in this area' are important issues for all businesses to be aware of. We have asked for feedback, and throughout the Report give readers the opportunity for dialogue with us. There are links to our website and those of other organisations and businesses interested in promoting the concept of triple bottom-line reporting.

The report looks at seven key areas of social impact: reliability of supply, employment and employability, the environment, education, excellence in the region, health and safety, and extra needs. It also reports on the work employees do in the community and how business can influence public policy.

Business has a role not just to respond to government policy but to help set and shape the agenda.

Our first Report also highlights the areas of our business that need, as we call it, 'intensive care'. For instance we need to take more action to address the balance of employment in areas of disability and equal opportunities. We also need to set targets and not just measures. The Report includes real business measures, that are part of our business planning and measuring process, but it does not include targets for where we'd like to be.

We will continue to report on our social impact. We aim to use this Report and face-to-face meetings to account for our social progress as we continue our dialogue with stakeholders and our commitment to meet their expectations.

This year we achieved two major milestones on our road to becoming a more socially responsible company. Our customer-facing businesses have been awarded Charter Mark status. This is national recognition of our high level of customer service. And we have achieved Investor in People status. This is national recognition of how we invest in our people – our employees. One of its major benefits is that it allows us to compare ourselves against the best practice of other companies and to continually improve the way we invest in, and train, our employees.

These are just two publicly recognised measures that we are taking the right steps. We will continue to check our progress against best practice models and we will continue to work on the national stage to develop corporate best practice. We will actively work with others to help address social issues, recognising that we have a shared responsibility and that business should indeed care.

Sandra Palmer is Marketing Communications Director for United Utilities, responsible for developing and delivering award-winning marketing communications strategies for the Customer Sales business. Sandra joined United Utilities in 1995 from the public sector, where she held senior communications posts with government-sponsored environmental charity Tidy Britain Group and its Going for Green initiative; the British Film Institute and a regional development agency. Sandra has lectured on communications and corporate social responsibility at Ashridge Management College, Warwick Business School and the Department of Journalism at Central Lancashire University.

United Utilities
Dawson House
Great Sankey
Warrington
Cheshire WA5 3JG

e-mail: sandra.palmer@uuplc.co.uk

A FTSE100 company, **United Utilities** owns North West Water and Norweb, the regional water and electricity companies in the north west of England. It also owns Norweb Telecom, Vertex and builds and operates utility services worldwide. The company has a long-term commitment to corporate social responsibility and was recognised for such in 2000 when it was presented with the Business in the Community Award for Impact on Society Company of the Year, in conjunction with the *Financial Times* and the Department for Trade and Industry.

Full details of United Utilities commitment to corporate social responsibility can be found at the website: www.unitedutilities.com

List of Acronyms
and Abbreviations

AA 1000	AccountAbility framework of tools to underpin quality in social and ethical accounting, auditing and reporting (SEA)
AAU	assigned amount unit
ABI	Association of British Insurers
ACCA	Association of Chartered Certified Accountants
ACBE	Advisory Committee on Business and the Environment
BAP	Biodiversity Action Plan (UK)
BAT	best available techniques
BREFs	BAT Reference Documents
BREMA	British Radio and Electronic Equipment Manufacturers' Association
BS 7750	British Standard Environmental Management System
BS 8800	Occupational Health and Safety Management System (guidance standard)
BSE	bovine spongiform encephalopathy
BSI	British Standards Institution
BT	British Telecommunications
BWEA	British Wind Energy Association
CAS	Corporate Advisory Services (ERM)
CBD	Convention on Biological Diversity
CBI	Confederation of British Industry
CDM	clean development mechanism
CEPAA	Council on Economic Priorities Accreditation Agency
CER	corporate environmental report
CHP	combined heat and power
CIA	Chemical Industries Association
CIMAH	Control of Industrial Major Accident Hazards
COMAH	Control of Major Accident Hazards

CJD	Creutzfeld–Jakob disease
CLRTAP	Control of Long-Range Transboundary Air Pollution
DETR	Department of the Environment, Transport and the Regions
DG ENV	Environment Directorate-General (European Commission), *formerly* DGXI
DGXI	*former* Directorate-General Environment, Nuclear Safety and Civil Protection (European Commission)
DNA	deoxyribonucleic acid
DTI	Department of Trade and Industry
DVD	digital versatile disc
EA	Environment Agency
EBRAR	Environmental and Biodiversity Risk Assessment Register
EEBPP	Energy Efficiency Best Practice Programme
EHS	environment, health and safety
EMAS	Eco-Management and Audit Scheme (European Commission)
EMAS II	Eco-Management and Audit Scheme version II (European Commission)
EMS	environmental management system
eol	end-of-life
EPE	environmental performance evaluation
ERA	Environmental Reporting Award
ERM	Environmental Resources Management
ERU	emission reduction unit
ETBPP	Environmental Technology Best Practice Programme
EU	European Union
EVA	ethylene vinyl acetate copolymer
FCCC	Framework Convention on Climate Change
FTSE100	*Financial Times* Stock Exchange listing of top 100 companies
GHGs	greenhouse gases
GM	genetically modified
GNP	gross national product
H&S	health and safety
HASAWA	Health and Safety at Work Act
HFC	hydrofluorocarbon
HSE	Health and Safety Executive
HVLP	high volume low pressure
ICT	information and communications technology
ILM	Integrated Logistics Management
IPC (IPPC)	Integrated Pollution Control (IPPC) (EU Directive)
IPP	Integrated Product Policy
IPPC	Integrated Pollution Prevention Control
ISDN	integrated services digital network
ISEA	Institute of Social and Ethical Accountability
ISO 14001	International Standard Environment Management Systems
ISO 14021	International Standard Environmental Labels and Declarations

ISO 14031	International Standard Environmental Performance Evaluation (guidelines)
ISO 9000	International Standard Quality Management Systems
ISO 9001	International Standard Quality Management Systems (requirements)
IT	information technology
LAs	local authorities
MACC2	Make a Corporate Commitment
MACs	marginal abatement costs
MEI	Matsushita Electric Industrial Co Ltd (Japan)
NECD	National Emissions Ceilings Directive
NFFO	Non Fossil Fuel Obligation
NGO	non-governmental organisation
NHBC	National House-Building Council
NIRO	Northern Ireland Renewables Obligation
NOx	nitrogen oxides
N/SVQ	National/Scottish Voluntary Qualifications
OH&S	occupational health and safety
OHS&E	occupational health and safety and environment
OHSAS 18001	Occupational Health and Safety Management Systems (specification)
P&G	Procter & Gamble
PBIT	profit before interest and taxation
PIRC	Pensions Industry Research Council
PLA	polylactide
PR	public relations
PRIMER	Producers' Institute for Management of Electronics Recycling
PRN	Packaging Recovery Notes
PUK	Panasonic UK Ltd
PWR	Packaging Waste Regulations
R&D	research and development
RCMS	responsible care management system
RDC	Regional Distribution Centre
SA 8000	Social Accountability 8000 (produced by CEPAA)
S&P	Standard & Poor's
SMART	**S**pecific, **M**easurable, **A**chievable, **R**ealistic, **T**ime-bound (of targets/objectives)
SRI	Socially Responsible Funds
SRO	Scottish Renewables Obligation
STEPS	Shell Tradable Emission Permit System
TUSDAC	Trade Unions and Sustainable Development Advisory Committee
UNEP	United Nations Environment Programme
UNICE	Union of Industrial and Employers' Confederations of Europe
WEEE	Waste Electrical and Electronic Equipment (EU Directive)

Useful Contacts

Andrew Dick
tel: +44 (0)20 7395 8054
e-mail: andrew.dick@cbi.org.uk

CBI Contour
Environmental Best Practice
Environmental Reporting
Management Systems and standards

Charlotte Granville-West
tel: +44 (0)20 7395 8053
e-mail: charlotte.granville-
west@cbi.org.uk

Air
Biodiversity
Environment Agency
IPPC
Risk assessment
Water

Richard Jackson
tel: +44 (0)20 7395 8052
e-mail: richard.jackson@cbi.org.uk

Climate Change
Climate Change Levy
Economic Instruments
Emissions Trading
Energy

Kamini Paul
tel: +44 (0)20 7395 8065
e-mail: kamini.paul@cbi.org.uk

Environmental Liability
IPP
Packaging
Producer responsibility
Waste
WEEE

Index of Advertisers

Other Earthscan Titles of Interest

The Link Between Company Environmental and Financial Performance
David Edwards, Andersen Consulting
£45.00 *paperback*   ISBN 1 85383 549 8

Environmental Risk Management – 2nd Edition
Paul Pritchard, Royal & SunAlliance
£45.00 *paperback*   ISBN 1 85383 598 6

Taking Responsibility
Personal Liability Under Environmental Law
Stephen Tromans and Justine Thornton
£45.00 *paperback*   ISBN 1 85383 597 8

Natural Capitalism
The Next Industrial Revolution
Paul Hawken, Amory B Lovins and L Hunter Lovins
£12.99 *paperback*   ISBN 1 85383 763 6

Installing Environmental Management Systems – Revised Edition
A Step-by-Step Guide
Christopher Sheldon and Mark Yoxon
£29.95 *paperback*   ISBN 1 85383 868 3
Forthcoming autumn 2001

Contaminated Land – 2nd Edition
Managing Legal Liabilities
Eversheds
£45.00 *paperback*   ISBN 185383 747 4

The Green Office Manual – 2nd Edition
A Guide to Responsible Practice
Wastebusters Ltd
£40.00 *paperback*   ISBN 1 85383 679 6

The Earthscan Reader in Business and Sustainable Development
Edited by Richard Starkey and Richard Welford
£19.95 *paperback*   ISBN 1 85383 639 7

Greening the Corporation
Management Strategy and the Environmental Challenge
Peter Thayer Robbins
£16.95 *paperback*   ISBN 1 85383 772 5
£45.00 *hardback*   ISBN 1 85383 771 7
Forthcoming autumn 2001

For more details visit our website

www.earthscan.co.uk